THE PLANT COMMUNITY
AS A WORKING MECHANISM

THE PLANT COMMUNITY
AS A WORKING MECHANISM

SPECIAL PUBLICATION NUMBER 1 OF THE
BRITISH ECOLOGICAL SOCIETY
PRODUCED AS A TRIBUTE TO

A. S. WATT

EDITED BY

E. I. NEWMAN

Department of Botany
University of Bristol
Bristol BS8 1UG, U.K.

BLACKWELL SCIENTIFIC PUBLICATIONS

OXFORD LONDON EDINBURGH
BOSTON MELBOURNE

1982

© 1982 by The British Ecological Society
and published for them by
Blackwell Scientific Publications
Osney Mead, Oxford, OX2 0EL
8 John Street, London, WC1N 2ES
9 Forrest Road, Edinburgh, EH1 2QH
52 Beacon Street, Boston
     Massachusetts 02108, U.S.A.
99 Barry Street, Carlton,
     Victoria 3053, Australia

First published 1982

Printed and bound in
Great Britain by
Spottiswoode Ballantyne Ltd
Colchester and London

DISTRIBUTORS

U.S.A.
   Blackwell Mosby Book Distributors
   11830 Westline Industrial Drive
   St Louis, Missouri 63141

Canada
   Blackwell Mosby Book Distributors
   120 Melford Drive, Scarborough
   Ontario, M1B 2X4

Australia
   Blackwell Scientific Book Distributors
   214 Berkeley Street, Carlton
   Victoria 3053

British Library
Cataloguing in Publication Data

The Plant community as a working
   mechanism.—(A Special publication
   of the British Ecological Society; I).
   1. Plant communities—Congresses
   2. Watt, A. S.   I. Newman, E. I.
   II. Watt, A. S.   III. Series   581.5′247
   QK911

ISBN 0-632-00839-3

CONTENTS

PREFACE

The title of this symposium is taken from a section heading in what is probably one of the most influential papers ever published on an ecological subject, A. S. Watt's (1947) paper on 'Pattern and process in the plant community'. The phrase 'the plant community as a working mechanism' summarizes the theme of this volume and of much of Watt's research, the attempt to understand how the composition, structure and dynamics of plant communities are controlled, in terms of the relations of the individual species to their environment and to each other.

The Council of the British Ecological Society decided that the Society should use the opportunity of Dr Watt's 90th birthday in 1982 to pay tribute to his outstanding contributions to ecology. Since he has published many important papers in the *Journal of Ecology*, over a span of more than 60 years, it seemed appropriate to honour him by a publication associated with the *Journal of Ecology*. On the following two pages P. Greig-Smith summarizes briefly Dr Watt's life and contribution to ecology. Some of the authors of the chapters of the symposium have, like Professor Greig-Smith, had the benefit of Dr Watt's teaching, either as undergraduates or research students, but they were not chosen primarily for that reason: the authors and topics have been chosen to illustrate active thinking and research on subjects akin to Dr Watt's own prevailing interests.

Although the authors were given considerable freedom to develop their subject matter in their own ways, the seven chapters are inter-related. Harper introduces the symposium, and relates it to Watt's own work, taking a critical look at plant ecology today and some of its weaknesses. Walker and Whitmore are both concerned with disturbance: Walker draws together evidence on the effects of fire over time scales ranging from years to millennia, and Whitmore considers the forest growth cycle in relation to gap phase regeneration. Regeneration is also a theme of my own paper, which uses models to predict how different species need to be in regeneration characteristics in order to coexist. Interest in mechanisms of coexistence also underlies the paper by Grubb, Kelly & Mitchley, who consider why some species are consistently commoner than others. Burdon summarizes evidence on the influence of pathogens on plant populations and the balance between species in communities, a subject which has not, up to now, received the attention it deserves. Finally, Ashton & Willis draw the symposium together with a detailed case study which shows how many of the factors considered separately in the other chapters interact in determining the success or failure of a single species, *Eucalyptus regnans*.

Every chapter of this symposium addresses questions which A. S. Watt has asked, and often begun to answer, during his long scientific career. This symposium is an indication both of his great influence on ecology and of how far-sighted and penetrating are the sorts of questions he has asked.

E. I. NEWMAN

REFERENCE

Watt, A. S. (1947). Pattern and process in the plant community. *Journal of Ecology*, **35**, 1–22.

A. S. Watt

A. S. WATT, F.R.S.: A BIOGRAPHICAL NOTE

Alexander Stuart Watt, now in his ninetieth year, is the son of an Aberdeenshire farmer. After schooling at Turriff Secondary School and Robert Gordon College, Aberdeen, he entered the University of Aberdeen, graduating M.A. and B.Sc.(Agric.) in 1913. In 1914 he went to Cambridge to work on oakwoods under A. G. (later Sir Arthur) Tansley, obtaining the B.A. by research (after interruption by military service from 1916 to 1918) in 1919. He had in the meantime been appointed Lecturer in Forest Botany and Forest Zoology at Aberdeen in 1915. While holding this post he continued research for the newly instituted Ph.D. of the University of Cambridge, working in the beechwoods of the south of England during vacations, and obtained the degree in 1924. In 1929 he accepted an invitation to become Gurney Lecturer in Forestry at Cambridge. With the demise of forestry as an undergraduate subject at Cambridge in 1933 he moved to the Botany School as Lecturer in Forest Botany—a title which scarcely reflected his wide interest in and influence on plant ecology—and remained in that post until his retirement in 1959. In 1950 he travelled widely in Australia, an experience he clearly found stimulating (1954*). His distinction as an ecologist was recognized in 1957 by election to the Royal Society.

Retirement was only nominal. He has continued to work actively in the field, and with recent papers in the *Journal of Ecology* (1981) has brought his span of publications in the Journal to 63 years. He was Visiting Lecturer at the University of Colorado in 1963 and, at an age when most Europeans would be deterred by the climate of Sudan, Visiting Professor at the University of Khartoum in 1965.

Watt has contributed much to the British Ecological Society. A member of the Council for many years, he was Honorary Treasurer from 1938 to 1949, Vice-President 1940–41 and President 1946–47. In 1970 he was one of the organizers of the successful symposium on 'The Scientific Management of Animal and Plant Communities for Conservation'.

Watt's earlier research was concerned with woodlands, oakwood (1919), beechwood (1923–34) and yew communities (1926), but his later work has centred on the Breckland, an area of low rainfall and predominantly sandy soils, conveniently near to Cambridge. The Breckland research, which exemplifies the value of working on whatever vegetation is at hand to investigate general ecological questions, falls into three parts: the factors determining the main vegetation types and their composition (1936–40, 1955), the details of change with time, particularly in relation to grazing by rabbits and its prevention (1957–62, 1974, 1981) and the ecology of bracken (1940–74, 1955, 1964). Latterly, his deep understanding of vegetation has been applied to problems of conservation (1971a, b).

It is an impressive record of research, the more so that it all rests on his own painstaking work in the field. Diverse though the vegetation is in which he has worked, a thread runs through his research, a compelling interest in the interaction between species in the community and in the differing potential of individuals of different ages, which limits the value of description of vegetation based only on records made at one time. This theme was developed in his Presidential Address to the British Ecological Society, 'Pattern and process in the plant community' (1947), expounding the view of the community as a mosaic of phases differing in the stage of the life cycle of the dominant species, with correlated effects on the accompanying species. This paper is still widely quoted and the

* Dates refer to Watt's publications. Other than those listed below all those quoted are in *Journal of Ecology*.

influence of its ideas can be seen in much contemporary work even where it is not cited directly. Two quotations illustrate his viewpoint.

'It is now half a century since the study of ecology was injected with the dynamic concept, yet in the vast output of literature stimulated by it there is no record of an attempt to apply dynamic principles to the elucidation of the plant community itself and to formulate laws according to which it maintains and regenerates itself' (1947).

'The quantitative assessment of the components of the plant community leaves you with a pile of bricks, timber and chimney pots without showing how they are arranged—and even when this is done it is a static picture that is presented. Now, whatever criticism one may have of Clements' notions of the climax, his injection of the dynamic principle into a static discipline did for ecology what Lyell did for geology, Darwin for biology and Dokuchaiv for soil science—he brought new life into it' (1961).

Not only was his approach original, though it is now very widely accepted; the details of his work also set an example. His initial work on woodland, with its emphasis on following the fate of a crop of seeds, can be seen as one of the earliest examples of the approach now recognized as population biology of plants. Unlike much of the earlier quantitative description of vegetation, his careful enumeration is a tool, of value in elucidating change, not an end in itself. In the assessment of his data, he was among the first to use statistical analysis, but it was applied only when it could be useful, never to cover up careless field-work nor to lend apparent erudition to the obvious. Not the least valuable lesson in the present climate of concern for quick results and early publication is the importance attached to long continued observation; many of his results are based on twenty years or more of continuous field-work, often with important conclusions not apparent until a late stage.

This brief note would be incomplete without some reference to Watt as a teacher. He eschewed flamboyance and playing to the gallery but, to those prepared to be interested, his lectures, like his writing, were lucid and inspiring; many who attended them must, like myself, look back on Watt's teaching, in the field as well as in the lecture theatre, as a major influence on their approach to ecology.

P. Greig-Smith
School of Plant Biology,
University College of North Wales,
Bangor, Gwynedd, LL57 2UW, U.K.

REFERENCES

Watt, A. S. (1940–71). Contributions to the ecology of bracken. *New Phytologist*, **39**, 401–422; **42**, 103–126; **44**, 156–178; **46**, 97–121; **49**, 308–327; **53**, 117–130; **66**, 75–84; **68**, 841–859; **69**, 431–449; **70**, 967–986.

Watt, A. S. (1954). The integrity of the water factor. *Vegetatio*, **5–6**, 29–35.

Watt, A. S. (1961). Ecology. *Contemporary Botanical Thought* (Ed. by A. M. MacLeod & L. S. Cobley), pp. 115–131. Oliver and Boyd, Edinburgh.

Watt, A. S. (1971a). Rare species in Breckland: their management for survival. *Journal of Applied Ecology*, **8**, 593–609.

Watt, A. S. (1971b). Factors controlling the floristic composition of some plant communities in Breckland. *The Scientific Management of Animal and Plant Communities for Conservation* (Ed. by E. Duffey & A. S. Watt), pp. 137–152. Blackwell Scientific Publications, Oxford.

Watt, A. S. (1974). Senescence and rejuvenation in ungrazed chalk grassland (grassland B) in Breckland; the significance of litter and moles. *Journal of Applied Ecology*. **11**, 1157–1171.

AFTER DESCRIPTION

JOHN L. HARPER

School of Plant Biology, University College of North Wales, Bangor, Gwynedd, LL57 2UW, U.K.

SUMMARY

In this paper I discuss what appear to me to be dangers in recent thinking and writing by ecologists.

Plant ecology has been dominated until recently by description of vegetation and analysis of the biology of single species. Both types of study lead to predictive statements—predicting where we will find particular vegetation (or species) and what sorts of behaviour (or form) we will find among the organisms in a particular region. Almost inevitably such descriptions use the species as the basis for description—yet it is far from clear that the conservative and stable characters and the breeding isolation that may be used to define such taxa are appropriate to define ecological units—knowing as we do the wide range of ecologically different behaviours that are included within single species. The most immediately relevant ecological differences between organisms may often be those involving intraspecific variation between (ecotypic differentiation) or within populations (polymorphisms). We lack an appropriate taxonomy to handle such problems—but the absence of tools does not mean that there is no problem.

The explanation for the behaviour of a particular organism, both its distribution and its physiology, may be explained in proximal terms—how its present properties explain what it now does and where it lives, *or* in ultimate terms—how it has come to possess its present properties and distribution. These two levels of explanation often become confused—particularly in the use of the word 'adaptation'. The word implies a teleology—that the organism is goal-seeking and its evolution has had goals and ends. It also implies that in some way the organism has gained a fit to its environment and that for many ecologists the task before them is to demonstrate how this fit works. It is suggested that most evolutionary processes lead to a narrowing and specialization such that restrictions are placed on what the organism can do and where it can live. What we see proximally are the consequences of such limitations and we might, with profit, change the nature of our question to take the form—what are the limitations in the form and behaviour of organisms that account for their present highly restricted distributions and behaviour? The ultimate question then becomes, how have such limitations arisen? A variety of forces, of which natural selection is only one (*perhaps* the most important) have to be taken into account as possibly restricting the range of species, forms and activities that have evolved and survived. The optimism of simple adaptive explanations is suspect.

Ultimate 'explanation' of the present behaviour and distribution of organisms depends on evolutionary speculation based on proximal observation of evolutionary forces in action; we assume that what we are now seeing in a short time is essentially what has continued over long periods. If this is so, ultimate ecological explanation has to be focused on the nature of present evolutionary processes in action, which must imply the study of genetic individuals and their descendants. It is questionable whether study at the community or habitat level, or holist studies of area performance can approach these problems. The case is argued for a concentration of effort on the lives and deaths of individual plants—a reductionist approach—as the most likely to reveal those forces at present operating to determine the distribution and abundance of plants, and those most likely to hint at the evolutionary forces that have left most present forms in their narrowly limited ruts of specialization.

0262–7027/82/0300–0011$02.00 © 1982 British Ecological Society

INTRODUCTION

In this paper I attempt to place the ecological studies of A. S. Watt against a background of what appear to me to be dangers in much ecological writing by other authors. Excessive preoccupation with the distribution of taxa (cartography for taxonomists), confusion about the compromises that have to be made between generality, precision and realism in ecological science, facile adaptationist and holist interpretation and the loose use of language appear to me to be especial dangers in the development of our science. The work of A. S. Watt offers model examples of ecological investigation and writing that is free from these dangers.

Much of the activity devoted to plant ecology, since the development of the subject as a science in its own right, has been essentially descriptive. It is natural that the first stages in the growth of any science (physical or biological) should consist of the description and ordering of the material for study. The next stage is to search for correlations between and causation of what has been described. In plant ecology the procedures for providing description have concentrated at two distinct levels—that of vegetation and that of the intimate study of the biology of individual species. The description of vegetation, whether in the hands of continental phytogeographers (Braun-Blanquet, Tüxen, etc.), the British tradition of Tansley, the American schools of Cain, Whittaker, Curtis, or the Russian schools of Sukatschev and his successors, has almost always been in terms of species composition. Vegetation is defined, whether objectively or subjectively, as assemblages of species which are treated as objects for classification or ordination and may be used in the construction of maps. In the ordering of vegetation types, other features of the environment (soil, climate, etc.) may either be used as sources of information with which the vegetation can be correlated (the one to predict the other) *or* combined with information about the species to give an ordering or classification, not only of the vegetation, but of habitat-environment complexes and ecosystems.

In contrast, in autecological description (e.g. in the Biological Flora of the British Isles) the aim has been to produce monographic treatments of individual species, their form, behaviour, distribution, and response to environmental factors such as frost, drought, soil nutrients, pathogens, predators, etc. Again the emphasis is on species; the taxonomic bias pervades almost all of descriptive ecology. These two broad categories of description give ecologists the equivalent of the telephone directory and 'Yellow Pages', the one describing who we can find where and the other describing who (plant or community) does what (and again where we can find him).

TAXONOMIC CHARACTERS AND ECOLOGICAL VARIATION

When we describe a species as having a particular distribution or particular ecological attributes we make a statement about a taxonomist's unit. Unfortunately for the ecologist, the criteria used by the taxonomist for the delineation of taxa are chosen deliberately from the conservative and stable features of morphology that are not subject to marked genetic variation, polymorphism or phenotypic change. These same criteria that are appropriate for the taxonomist may be quite inappropriate for describing the ecologically relevant differences between individuals, populations and communities. The problem is illustrated by considering the meaning to be attached to the statement that a particular species has a 'wide distribution'. This may mean either (a) that the individuals of the species, though genetically narrowly based, have wide tolerances or plasticity so that individuals sampled over the range of the species will behave equally well over that range, and are mutually exchangeable (Baker (1965) considers that many weeds have such all-purpose genotypes); or (b) that the individuals have very narrow tolerances but the nature of the species is such that the taxonomist includes a wide range of locally specialized genotypes within one taxon. In the first case information about the distribution of the taxon tells us something about the physiology of individuals and in the second case about genetic variation within and between populations.

The second class of species is illustrated by the work of Bradshaw and his colleagues (Antonovics, Bradshaw & Turner 1971) on *Agrostis tenuis* of which local populations on metal mines may be tolerant of zinc, copper, lead or other toxic minerals; the tolerances are specific. The species, *Agrostis tenuis*, has a wide distribution, but the populations are locally specialized and apparently incapable of persisting in each others' habitat. Individuals are not interchangeable over the range of the species. When we find that individuals of a species are not interchangeable between places in its distribution, the statement that the *species* has a particular distribution tells the taxonomist where he could find a plant to which he would give that name, but it does not tell the ecologist which plants will live where. The very extensive studies of Turesson, Gregor, Clausen, Keck, Hiesey, Bradshaw, and others have shown that plants of a single species sampled from a wide range of habitats and grown together in an experimental garden differ, often profoundly, in features of growth, form and life cycle. More recently Clegg (1978) and Turkington & Harper (1979) have made reciprocal transplant experiments which demonstrate that locally differentiated populations occur within small, superficially uniform areas of vegetation and, more significant, that such plants of the same species are often not mutually interchangeable. Not only are individuals within a local population different, but the differences are ecologically relevant, affecting fitness attributes. Such fine scale biotic differentiation as occurs between clones of *Trifolium repens* within one small pasture (1 ha) (Turkington & Harper 1979) and between populations of *Ranunculus repens* across woodland/grassland borders over distances of less than 50 m (Clegg 1978), emphasizes the impossibility of making predictive statements, about what genotypes will grow where, over even small distances. The same point is nicely made by Haukioja's (1980) study of the distribution of *Betula pubescens* in Finland. This species has a wide distribution from north to south but reciprocal plantings, made between northern and southern populations, showed striking differences in the damage done by insect predators (*Oporinia*) to the plants growing in their alien and natural environments. The same sorts of differences were found between populations on the slopes of a single fjell over an altitude gradient 80–330 m above sea level. The distribution map of *Betula pubescens* in Finland tells us where a taxonomist could expect to find a plant to which he could give that name—it does not in any sense tell the ecologist which plants bearing that name would suffer heavy insect defoliation in different parts of that distribution or which would live or die.

The failure of taxonomic categories to fit as ecological categories is not surprising when it is remembered that the taxonomist searches for stable, conservative characters for his groupings, yet it may be just the taxonomically useless characters that are mainly responsible for determining the precise ecologies of organisms. Properties such as the degree of plasticity, germination time, and form of the whole plant can clearly be critical in determining the life or death of individuals and so contribute to their fitness. Similarly, the position that a plant comes to occupy in a hierarchy of competing neighbours of its own and other species depends greatly on its branching form and stature, yet it is the shape of the conservative organs, the bits, such as leaves, petals, etc., not the form of the whole, on which the taxonomy is usually based. It may be that it is just those differences between plants that the taxonomist avoids, which the ecologist needs most strongly for effective description.

The ecological significance of intraspecific variation is often displayed where populations of a species are polymorphic and so individuals with different biologies are found intermingled within a single habitat. Even the sexual dimorphism of dioecious species is such that plants of the two sexes may play quite different roles within the community. The vegetative precocity of males compared to females in *Rumex acetosella* (Putwain & Harper 1972) and in *Spinacia oleracea* (Onyekwelu & Harper 1979) illustrate such ecological differentiation. Ecological differences between the sexes may extend to a geographic scale—males of *Petasites hybridus* are locally common throughout the British Isles, but females have a narrowly limited distribution (Clapham, Tutin & Warburg 1962).

A variety of ecological studies have been made of cyanogenesis polymorphism in *Lotus corniculatus* (Jones 1962) and *Trifolium repens* (Dirzo & Harper 1982a, b). The expression of cyanogenesis is under genetic control at just two loci; if cyanogenesis is expressed the plant is almost wholly protected against being eaten by slugs or snails. If slugs and snails are abundant in a pasture this simple genetic difference may determine whether individual plants will live or die and such intraspecific variation may be at least as important in the ecology of *Lotus* and *Trifolium* as the presence or absence of other plant species in the sward.

The extent of ecologically relevant polymorphism in plant populations within small areas is emphasized by Burdon (1980) in his study of *Trifolium repens*. In the 1-ha field of permanent grassland intensively studied at Aber, near Bangor, North Wales, and already referred to in this paper, *T. repens* is abundant. Fifty clones, sampled from a grid spaced across the field, differed from each other on average with respect to 3·3 non-flowering characters, apparently all genetically controlled. One pair of clones differed in thirteen statistically significant respects! The characters included major aspects of growth such as relative growth rate, leaf area, petiole length, resistance or susceptibility to two pathogens, cyanogenesis, leaf marks and other characters. Each of these characters had been shown to be of selective importance by one author or another in *T. repens* or some other species of *Trifolium*. A descriptive ecology that simply records *T. repens* as present, its abundance, its microdistribution, its relationship to other species and to physical factors hides all such information. Any attempt to progress from a descriptive ecology that describes where species are found to an interpretative ecology that accounts for the distributions should ask, in each case, how far the taxonomist's species represents an ecological unity or a compendium of significantly different ecologies. Problems of the distribution and abundance of species may need to be seen as essentially problems in genetics!

The description of the autecology of plants is beset with many of the same problems—not least (following the above arguments) that the ecologically relevant biology of an organism is not defined by attaching a binomial. Even supposedly fundamental physiological attributes such as the relative growth rate may be genetically variable within a small local population (Burdon & Harper 1980).

Autecological studies may be expected to provide an accumulation of data for a matrix from which we might extract significant ecological generalizations. It is an expectation that the variety of species and forms present in a particular type of habitat will have features in common. We expect that the plants that are found in, say, a waterlogged habitat will all possess properties that enable them to live in a waterlogged habitat and we can then use these similarities as a demonstration of the powerful forces of evolution and ecology that constrain the range of forms found in a particular environment. However, it is also a reasonable expectation that species that live together in a habitat will differ. It is part of the conventional wisdom of theoretical ecology (Gause's hypothesis) that two or more species that persist together in a habitat without one succeeding at the expense of another must differ in the ways that they exploit that habitat—differing in such a manner that each suffers more from its own increasing density than from that of its neighbours. (This problem is discussed for plants by Harper *et al.* (1961) and Newman (this volume).) We thus have two contrasting expectations. If all the plants in a waterlogged habitat contain aerenchyma this can be seen (and taught!) as a splendid example of convergent evolution; if some have aerenchyma, some have superficial roots, others have mechanisms that prevent the formation of toxic anaerobic byproducts and yet others are able to metabolize such products, we have a splendid example of evolutionary divergence, a variety of 'solutions' to a single environmental 'problem'. Thus, if the biologist finds similarities between organisms in a habitat, he can feel satisfied that he has an ecological convergence and he can find equal satisfaction in demonstrating differences that illustrate necessary ecological divergence. There are no losers in this type of investigation—theory is so broad that every observation can be fitted. Both similarities and differences can be accounted for as 'adaptations' to the environment ('Eureka' ecology!).

ADAPTATION, STRATEGY AND STRESS

The concept of 'adaptation' is fundamental to most ecologists who seek to extend descriptive syn- or autecology to invoke causal (as opposed to purely correlative) explanations of ecological phenomena. Organisms are thought of as fitted in a precise lock/key relationship with both the physical and biotic factors of the environment. If an organism is found more or less reliably in a particular type of environment it is easy to take this as a demonstration that it has, in some way, been programmed to that end. It then remains for the ecologist to discover and describe those properties that confer this precise fit. Darwin (1859) carefully denied that the process of natural selection was the sole cause of the evolutionary process—yet post-Darwinian Victorian optimism continues in much ecological thinking—that the organism should and can be interpreted as a perfected product of an all embracing, idealizing and optimizing process of natural selection. Gould & Lewontin (1979) have cogently and wittily criticized this Panglossian paradigm in evolutionary theory. The organisms that we study behave and live where they do because of properties passed to them by their ancestors. Some ancestors have left more descendants than others so the nature of populations, species and communities has changed; they have evolved. A variety of forces influence which organisms (or genes) leave descendants and so determine their behaviour. There is much more to these processes than the naive view of all powerful natural selection constantly moulding and idealizing each population towards optimal behaviour in a preferred niche. Factors that seem likely to have determined which individuals have contributed the descendants that now make up our flora (and fauna) include the following seven.

(1) *Founder effects.* When new populations establish after catastrophes or invasions of new areas the gene pool represented in the founders may be very limited and hence chance elements may play a role in the direction of subsequent evolutionary change.

(2) *Archetype effects.* No evolutionary process starts with a fresh sheet: always the

process acts on ancestors that are more or less complex organized systems, and there are therefore limits on what new changes are possible. It is for this reason that Jacob (1977) has described the evolutionary process as 'tinkering'—existing systems are altered, patched, twisted and refitted, but always carry some of the imprint and limits imposed by the nature of the original. (It is relatively easy to tinker a saucepan, but not a bicycle, from an old kettle.) Stebbins (1971) considers the effect that the same selective force acting to favour increased fecundity might have, when applied to a population of lilies and of a grass. The archetypic constraints within the Gramineae make it exceedingly unlikely that the grass will respond by increasing the number of ovules per ovary and more likely that the response will be to increase the number of ovaries per plant. In contrast, the lily with already multi-ovuled ovaries, may be more likely to respond to selection by increasing the number of ovules per ovary. In each case the direction that an evolutionary pathway takes under selection will be under archetypic constraints. The response to selection follows the easy line of least resistance, not necessarily the optimal route (Stebbins (1970) has called this process 'selection along the line of least resistance').

(3) *Available genetic resources.* Different populations and species contain different and limited resources of variation. The direction of an evolutionary pathway under a particular selection pressure will be determined in part by which mutations and gene recombinants are exposed in the population. Resistance to triazine herbicides has developed in *Poa annua*, *Senecio vulgaris* and *Chenopodium album* (see, for example, Holliday & Putwain 1977). These triazine-tolerant forms, but few other species, now colonize some agricultural habitats in which these herbicides are regularly used. Presumably, this reflects the difference in the nature of the genetic variability available within populations of the different weed species at the time at which the selection process acted. It is not a necessary consequence of the application of a selective process that a population will respond by genetic change in the direction that minimizes the effect of the selection. The direction of evolutionary change in a population (if any) is constrained within the limits of its genetics. Consequently, we cannot argue in reverse that those features that we observe in an organism in nature are, in any sense, optimal solutions to past selective forces.

(4) *Pleiotropism and linkage disequilibrium.* Both of these phenomena may result in diverse properties of an organism being inherited together, with the result that one set of properties may be carried in the evolutionary process on the heels of another.

(5) *Allometry.* Gould & Lewontin (1979) emphasize particularly the role in the evolutionary process of allometries in which groups of apparently independent characters reflect a particular aspect of growth expressed in a variety of different organs. The phenomenon of gigantism is, for example, rarely confined to one organ on a plant but will normally be expressed in a large number of manifestations of growth.

(6) *Selection.* This implies a non-random (i.e. directional, stabilizing or disruptive) change in the genetic composition of a population and, of all the forces of evolution, is the one conventionally thought of (and probably rightly) as the prime cause. It may take the form of a process that leads towards some asymptotic condition in which members of a population become, over generations, tolerant of the major repeated hazards, particularly physical hazards, in the environment (e.g. drought, frost). However, selection may also be exercised by biotic forces, such as inter- and intraspecific competition from neighbours or predation and parasitism. In such cases the reaction to selection may not be asymptotic because each change in predator or prey or in one of a pair of competitors produces changes in the selective forces acting on the other which then exerts reciprocal selection. Such reciprocating selection has no obvious end point and the process of coevolution that results may continue until only the availability of further genetic variation places limits on

evolutionary change. The coevolutionary process is most beautifully illustrated in the breeding of crop varieties by man where it has become necessary continually to develop new strains of crop to combat the apparently endless coevolution of new strains of pathogens. The accumulation of batteries of alkaloids within individual species of *Senecio* (Barger & Blackie 1937) or of *Lupinus* (Dolinger *et al.* 1973) may represent a comparable phenomenon in nature involving an endlessly reciprocating selection between coevolving host plants and their predators. Where selection is biotic, so that coevolutionary processes may occur, the consequence is likely to be the accumulation of complexity; the interacting populations in the coevolutionary process each drive the other further into an ever-deepening rut of specialization (Huffaker 1964). This is a very different situation from a precise fit between organism and environment.

(7) *Compensation.* Any evolved change in part of an organism's form or behaviour seems likely to be developed at the expense of compensating changes in some other aspects of the life of the same organism. An increased expenditure of resources on reproduction is likely to reduce the resources available for vegetative growth; an increase in seed size is likely to be compensated by a reduction in seed number. Increased expenditure of carbon or minerals on root growth is likely to limit the resources available for shoot growth. There is every reason to suppose that most aspects of the form and behaviour of an organism represent the result of some set of compromises. Thus it is dangerous to search for interpretations of any isolated part of the form or behaviour of an organism and to give it an explanation in its own right as an optimal system. It is particularly tempting to do this when teaching students in the field and to assign *ex cathedra* 'explanation' to every feature of every organism present in a habitat. Facile adaptationist guesswork used to explain everything that we observe in nature scarcely serves to make ecology an effective medium for teaching the principles of science.

The list of evolutionary forces above is certainly not exhaustive. They have in common that they are all likely to constrain and narrow the range of activities of organisms (though they may increase the variation between them). They are also likely to limit the range of habitats in which organisms may complete their life cycles and leave descendants. This means that rather than concentrating on a search for the ways in which organisms are perfectly suited to their environments, we might more healthily concentrate on the nature of the limitations that constrain where they live. We might usefully ask not what is it about an organism that enables it to live where it does, but what are the limits and constraints that prevent it living elsewhere? It is perhaps because we usually seek to explain the perfection of plants (as does the zoologist for his animals—see Cain (1964)), that so little plant ecology has been concerned to discover what goes wrong with plants when they are grown in *communities* outside their normal range. This would appear to be the ideal way to demonstrate the real extent and proximal cause of the narrow specialization of most plant forms. The fact that forms of so many species can be grown successfully in cultivation, in alien soils and climates far outside their natural distribution, suggests that much of the narrow ecological range of species in nature is determined proximally and perhaps also ultimately by the biotic rather than the physical and chemical forces in the environment. It is for this reason that we need many more experiments in which plants are introduced to alien communities where they may reveal just what are the real causes of their failure to maintain themselves.

The word 'adaptation' appears so often in the writings of ecologists (including some of my own) that its meaning bears some examination. Stern (1970) has reviewed some of the usages. There seem to be three common meanings.

(1) It may mean the change that occurs in a phenotype as a result of some

environmental experience where the change is *assumed* to improve the ability of the organism to continue growth (or better, its chance of leaving descendants) compared with an organism that did not undergo the change. Frost hardening is an example. It is important, however, to recognize that no assumptions can be made that any phenotypic change in a plant in response to the environment is in some way an improvement. Indeed, some phenotypic responses, such as the formation of a restricted superficial root system by plants which become waterlogged in spring, probably reduce the chance that the plant will complete its life cycle later. It is certainly an act of gross optimism to *assume* that all environmentally induced phenotypic change increases fitness—that all reactions to environmental change are in such a direction as to minimize its damaging consequences and maximize the advantages. When the word is invoked to describe phenotypic change, it needs very precise usage if it is not to be trite.

(2) An adaptation is any aspect of form or behaviour that *at a reasonable guess* is the result of natural selection (see Williams 1966). This is probably the commonest usage and assumes that of the incomplete list of evolutionary forces listed above, natural selection is so overwhelmingly the most important that we can assume it accounts for most of what we observe. It is doubtful whether we can ever make that assumption.

(3) An adaptation is any feature of form or behaviour that can, in retrospect, be called on to account for the ability of an organism to live where, or do what, it does. It is any feature of an organism that can be explained away as a 'good thing' in a perfect world: any feature that an intelligent creator might have provided had it been his intention that the plant was destined to live where it is now found.

I suspect that the word 'adaptation' has lost its value in biological science and now serves mainly to give superficial respectability to what is really intolerable teleology. The prefix *ad*, with its implication of *to* or *for* the future, enshrines the teleology; only deliberate substitution by phrases that say exactly what is meant in a given context, or new words, can rid ecological writing, and especially teaching, of the glib teleology that mocks the science. Perhaps *ab*aptation is a better word with the prefix *ab* now implying all those features of the organism's form and behaviour that characterize it by virtue of events *by*, *with* or *from* the past.

Ecological interpretation would gain in precision if every time an author used the word adaptation he forced himself to state precisely what he meant. In particular, this means differentiating between proximal and ultimate explanations for biological properties, i.e. differentiating between the behaviour of the organism as explained in terms of its present properties (proximal explanation) and the explanation of how it comes to possess such properties, i.e. the evolutionary forces that acted on the populations from which the organism's ancestors were drawn (ultimate explanation). This distinction becomes especially confused in the now common use of the word 'strategy' to describe the programmed biology, especially the life cycle, of an organism. The term 'strategy' sounds like a teleology—as if the organism has a planned campaign of behaviour aimed at the future. Ghiselin (1974) justifies the use of the word and puts it into terms appropriate for ultimate explanation: a strategy 'somewhat resembles the proverbial military school which produces officers admirably equipped to fight the battles of previous wars', i.e. a strategy is a property that is *by*, *with* or *from* the past. In most writing by ecologists, however, 'strategy' seems to imply a plan for future survival—a programme designed to achieve a goal.

Anthropomorphisms creep easily into science—perhaps because they are so often a means by which a teacher can create involvement and interest in a class of students.

Adaptation, Strategy and Stress are three emotive words that are now common in ecological writing. 'Stress' carries over easily from social and medical sciences to imply in biology 'what I don't think I would like if I was a (buttercup, kangaroo, flea, beetle, etc.)' Often the word is wholly redundant as in 'the effect of temperature stress' (= the effect of temperature), 'the effect of density stress' (= the effect of density) or 'the effect of water stress' (= the effect of drought or sometimes = the effect of waterlogging!). Sometimes the word is used to describe a force or a stimulus (drought, pollution, cold) and sometimes a reaction or response. Pickering (1961), a professor of medicine, wrote about the use of this word: '... stress, again an old word, but now part of the popular modes of expression in jargonese. I am never sure what it means. The classical language of biology uses two expressions—namely, *stimulus*, to describe a change in environment, and *response*, to describe the resulting change in the organism. The modern use of the word stress we owe to Selye, who uses it to express the first stage in the common reaction of the mammal to a variety of harmful environmental changes. It was thus the first stage in the *response* to certain kinds of stimuli.... Others quite clearly use the word to describe environmental changes, that is *stimuli*. For example a man is exposed to stress, or, in certain examples, to heat stress. This word is now in common use, and by the most distinguished scientists and laymen. Whenever I meet it, I set it aside, because frankly I do not know what it means and I fear it is, like shock, another of these words of deception. I find it difficult adequately to express my surprise and horror that contemporary science should tolerate this confusion of stimulus and response.'

PROBLEMS OF HOLISM AND REDUCTIONISM

If ecology is the study of the relationship between organisms and their environment, then agriculture and forest science are part of ecology—an applied sub-set of the science. However, in so far as ecology is an experimental science, the contribution from agriculture and forestry is much the largest part of the whole. For many ecologists the rightful field of study is in the description of the undisturbed wilderness of communities untouched by man—the natural communities of nature with their richness of biotic diversity within and between habitats. The special concern for the conservation of natural communities—as places in which the real world of nature can be admired and explored—sets some ecologists apart from those who study man-managed environments. There is an antithesis in the practice, and often antagonism between the practitioners, of ecology in the wilderness and ecology on the farm. This antithesis seems to have two elements. Firstly, land is a limited resource for which special interests compete—agriculture, forestry, conservation, recreation, roads, building, etc. There are expected tensions between these interests. It is healthy when ecology is less often identified with just one of these interests (e.g. conservation) and more often seen as a science underlying the study of all organisms in all environments (including man in factories and cities). Secondly, there has been a mainstream of holist philosophy among many ecologists—an almost religious view that natural communities of organisms have properties that are more than the sum of individual effects plus their interactions. To the holist, any man-made alteration of the whole is a sacrilege.

The holist attitude restricts the science that can usefully be practised to essentially descriptive and correlative activites. For those who make it an act of faith that the whole is more than the sum of its parts plus their interactions, the behaviour of deliberately simplified systems is irrelevant to understanding. Nowhere is this holism more apparent than in the way ecology is taught in many schools and universities with its emphasis on complex systems and the ways in which they can be described. There is probably no other science in which students are taught by being dropped into the deep end of complexity. Ecology is usually introduced to school children by showing them oakwoods, chalk grassland, an intertidal zone or a pond or lake margin and perhaps asking them to describe it. This is equivalent to introducing a child to chemistry by showing him the structure of haemoglobin or a DNA helix. The results of such holist teaching are to be seen in an emotive pseudo-science of ecology with its own language exemplified by a statement in the VIIth Report of the Royal Commission on Environmental Pollution: 'Until recently, agriculture has developed in harmony with the environment'. (What is harmony? How do we measure it?) If we accept (again as an act of faith) that the activities of communities of organisms are no more than the sum of the activities of their parts *plus* their interactions it becomes appropriate to break down the whole into the parts and study them separately. Subsequently, it should, again as an act of faith, be possible to reassemble the whole, stage by stage, and approach an understanding of its workings. This approach demands that we begin by looking at simple systems. It must mean looking at individual organisms and their behaviour alone and when brought together in the very simplest communities of single species in very simple physical conditions. From this stage, increasing complexity can be introduced step by step. To an ecologist who is not obsessed by holism it comes as a delightful surprise to realise that agricultural science (and to some extent forestry) has laid down just these basic elements of ecological science. Agriculture usually represents the growth of one or very few chosen species in simplified conditions (deliberately made uniform). It is indeed from agricultural science that the plant ecologist derives most of his basic information about:

(1) the effects of physical variables in the environment (temperature, nutrition, radiation, water supply) on plant growth;

(2) the effects of density on plant and crop performance (intraspecific competition);

(3) the effects of two species interactions at a single trophic level (weed-crop interactions, legume-grass interactions, the behaviour of crop mixtures);

(4) the effects of host-parasite interactions (crop diseases);

(5) the effects of predator-prey interactions (the grazing animal interacting with the grazing sward and the effects of pests on monocultures);

(6) the interactions of three species systems, plant-animal-animal (biological control of insect pests of crops) and plant-plant-animal (grazing effects on grass-legume pastures and the biological control of weeds by introduced insects);

(7) symbiosis (legume-*Rhizobium* associations);

(8) community assimilation, radiation capture and productivity;

(9) genetic variation within and between populations.

In agricultural science the workings of simple systems begin to be understood at a level that makes predictive ecology possible. A splendid examples lies in the management of ryegrass-white clover pastures where there is a highly sophisticated technology available for maintaining particular balances between grass and clover. Studies that developed particularly from the work of pioneer agronomists (e.g. Jones 1933) led to the remarkable situation in the 1950s in which various farming organizations were giving medals or silver cups and prizes to farmers who had achieved 'ideal' pasture composition. Careful balancing of nitrogen and phosphorus fertilizers and controlled grazing can produce a desired balance of species. There is probably no example among world vegetation types of a community so deeply understood and with its interactions so deeply researched (Wilson 1978). There is far to go before such a predictive ecology might be realized for the management of more complex vegetation. A single example illustrates the problem. The fritillary (*Fritillaria meleagris*) is a bulbous plant found very locally in Britain (and usually then in high abundance) in low lying hay meadows on neutral soil. The plant is beautiful in flower, spectacular en masse and the populations are cherished as objects for conservation. One fritillary dominated meadow (Magdalen Meadow) lies within the ground of Magdalen College, Oxford. It adjoins a once elm-wooded grove on which a herd of fallow deer has been maintained for many years. The deer carry a heavy parasite burden and should have a change of ground. The fritillary meadow is managed by cutting for hay after the fritillaries have flowered and set seed and is then grazed by bullocks. The 'simple' management question that has been posed to the ecologist is 'What would happen to the fritillaries if the deer were moved to the meadow?'. Such a problem has one simple holist answer—that the fritillary forms part of a system in which any change poses risks to the whole. Yet, deep understanding of the interaction of fritillaries with other plant species in the meadow, knowledge of the species' life cycle and population biology, its genetic variation and degree of local specialization, coupled with a knowledge of the differences between the grazing activities of bullocks and deer (dunging, urinating, trampling and defoliating), might allow a more subtle, useful and fundamental answer to the question. The holist answer is safe and ignorant. Presumably, this sort of question could be answered empirically on an agricultural analogy, by a simple randomized block experiment in the meadow involving deer *v*. bullocks, hay *v*. continuous grazing and coupled with a detailed analysis of changes in the flora of the plots. Such an experiment might, in time, give guidance for a changed programme of management of the meadow and of the deer. The results might, however, be locally specific and relevant only locally, unique to that meadow.

I presume that there is more to ecological science in the post-descriptive phase than acquiring the ability to handle unique, anecdotal management problems. I presume that it is a search for wider ecological generalizations that will dominate the post-descriptive phase in plant ecology. However, the search for generalities in ecology has been disappointing—more so in plant than in animal ecology. The few generalities that have emerged come from studies of stands of single species. The 3/2 thinning law defines the upper limits to the combination of sizes and numbers of plants that can persist in a monoculture (White & Harper 1970; Gorham 1979). This law has generality extending over more than eighty species ranging from *Sphagnum* to *Pinus* and from annual weeds to herbage grasses. Similarly, the law of constant final yield states that over a wide range of densities, yield per unit area becomes independent of the number of plants sown. This holds true for a great number of crop species and is probably a rather general law (White 1980).

PRECISION, REALISM AND GENERALITY

The search for generalities characterizes all sciences, and they usually take the form of established, reliable correlations or chains of causation. The ecologist sits on a trilemma in deciding at what level to search for correlation and causation. He may look for high precision, high realism or high generality or have to be prepared to compromise between these aims (Levins 1966). The search for precision may lead him to work with unique genotypes in controlled environments designed to minimize the background of environmental noise. The risks are then that his experimental results have no realism because natural populations are usually genotypically polymorphic and environments are normally noisy. He may emphasize realism by studying the behaviour of individual plants in the field; then only very large and extended studies may allow significant effects to be detected because of the magnitude of environmental noise and variance over space and time against which they must be compared. The search for generality may sacrifice both realism and precision and lead the ecologist to large-scale survey which runs the risk of yielding results that are only trite and superficial.

The research problem of deciding whether to place emphasis on precision, realism or generality is less in agriculture than in the ecology of natural systems. This is largely because it is in the nature of agricultural practice that the physical and biotic environments are simplified and variation from place to place is reduced by activities such as drainage or irrigation, fertilizer application, cultivation, pest and disease control and by the use of a limited number of crop varieties. Thus, local 'land' varieties of crops are replaced by more and more widely used specific cultivars or strains. In agricultural ecology precision, realism and generalization become more compatible in scientific enquiry. It may well be for this reason that much of the development of ecology as a science in the post-descriptive phase will come from the study of man-managed ecosystems.

There is, however, one important difference between the ways that we approach the ecology of agricultural and forestry systems and those of natural environments. The measure of the efficiency in agriculture and forestry is the profitability of areas of land or of populations of the crop. Here, there are no semantic objections to using all the teleological concepts of efficiency, goal and adaptation. A variety of a cereal crop *is* 'adapted' in the strict sense of being planned for the future. The ecology of an agricultural system is managed with a goal (profit), using organisms bred to specific ends. The efficiency of such a system can be measured and compared with others in respect of profit made, dry matter fixed, nutrients cycled and food produced but there are dangers when this type of thinking is transferred to the ecology of natural communities. In nature, few individual organisms respond to the area of the community in which they live. Moreover, the driving force in both succession and selection within a community is the determination of which individuals will leave descendants rather than some property of the group or community. Indeed, those features of individual ancestors that have led them to leave more descendants than their neighbours in nature may be quite irrelevant to performance per unit area.

It is instructive to consider just which sorts of heritable properties of individuals will be favoured in the struggle for existence. In higher plants these may include a growth habit that overlaps neighbours, depriving them of light; the quick capture of nutrient resources, even in excess of individual requirements, if capture deprives a neighbour of the same resources; rapid exhaustion of water supplies if this leaves a neighbour disproportionately deprived; retention of, rather than recycling of nutrients (if this deprives a neighbour). Indeed, unless there are strong processes of kin-selection operating amongst plant populations, we might expect to see emerging in nature, predominately those evolved properties that confer 'selfish' individualistic traits. There is no reason to expect that the evolution of organisms has been conditioned in nature towards any goal of productivity, stability or community yield, yet it is by just these criteria that we judge the efficiency of man-managed crops and forests. For these we breed varieties of plant for their population or group yield—considerations of individual fitness cease to have the relevance that they have in nature. If it is true that natural vegetation is what it is because of the evolution of individualistic selfish traits, it is unlikely that we will find ultimate explanations of ecological phenomena in nature at the area or population level. Both area and ecosystem are highly appropriate ecological units for the agronomist and forester but may be inappropriate levels of study at which to seek 'ultimate' explanations of ecological behaviour. If this is the case, only superficial generalities about ecosystems may be expected to emerge from systems studies of natural communities.

It is not the aim of this paper to be iconoclastic or defeatist, rather it is to set a background against which to comment on the significance of the contribution by A. S. Watt to the science of plant ecology. It is doubtful whether any other individual in the history of the science has contributed so much to the detailed description of the proximal events that influence the behaviour of individual plants in nature. The attention of his research, whether on beech woodland (e.g. Watt 1934), stands of bracken (e.g. Watt 1940, 1945) or in the patterned communities of the Breckland (e.g. Watt 1957) has, in every case, focused on the intimate, proximal events that affect individual plants and that, taken together, account for the nature of the communities he has studied. The reductionist focus on which seeds of the oak produce seedlings, which fronds of bracken become frost-damaged or which plants occupy a point in space in succeeding years in the Breckland, give us a description of ecological events that combine the high realism of a field observation with high precision: it is particularly characteristic of his studies that they extend over long periods of time so that the year to year variations in climate and other factors do not obscure general trends. The sacrifice in these studies is, of course, in generality. The precise behaviour of the populations of plants on Watt's Breckland plots may not be repeated anywhere else on the face of the earth. Genotypes of bracken may behave quite differently in other places and *Fagus sylvatica* may behave quite differently away from the Chilterns. Watt's contribution to ecological generality comes at a quite different level. It is in his ways of looking at natural systems and his concepts of plant community behaviour (regeneration cycles, pattern and process) that he introduces generalized attitudes that profoundly affect the work of others. His work is perhaps the model of a reductionism out of which conceptual generalizations emerge. His writing is remarkably free of any confusion between proximal and ultimate causes. What he discovers happening in his communities becomes the type of phenomenon that other people then search for in theirs. I suggest that it is from the repeated studies by *individuals* like Watt of *individual* plants in particular local conditions that most ecological generalizations are likely to emerge. Many of these are likely to be concepts, not laws. Further, I suggest that it is from the work of many individuals, working scattered over a variety of parts of the world, but concentrating their attention over long periods on the behaviour of individual plants, that the development of ecology as a generalizing and predictive science may become possible. I have little faith that either holists or the big science of the large team will be as effective.

It is greatly to be hoped that others will build upon the long extended field observations of A. S. Watt, particularly in the Breckland. If, in such communities, about which we now know so much, there can be some concentration of effort in exploring the genetics and breeding behaviour of the plants present, and some controlled perturbations of the vegetation that may reveal the nature of its controlling factors, we have a reasonable hope, not only of obtaining proximal interpretation of the ecology of the organisms but of making better guesses about that nature of ultimate causes. The detailed analysis of proximal ecological events is the only means by which we can reasonably hope to inform our guesses about the ultimate causes of the ways in which organisms behave.

ACKNOWLEDGMENT

I am grateful to Dr J. W. Tinsley for telling me of the reference to Sir George Pickering's paper.

REFERENCES

Antonovics, J., Bradshaw, A. D. & Turner, R. G. (1971). Heavy metal tolerance in plants. *Advances in Ecological Research*, **7**, 1–85.

Baker, H. G. (1965). Characteristics and modes of origin of weeds. *The Genetics of Colonizing Species* (Ed. by H. G. Baker & G. L. Stebbins), pp. 147–172. Academic Press, New York.

Barger, G. & Blackie, J. J. (1937). Alkaloids of *Senecio*. III. Jacobine, Jacodine and Jaconine. *Journal of the Chemical Society*, 1937, 584–586.

Burdon, J. J. (1980). Intra-specific diversity in a natural population of *Trifolium repens*. *Journal of Ecology*, **68**, 717–736.

Burdon, J. J. & Harper, J. L. (1980). Relative growth rates of individual members of a plant population. *Journal of Ecology*, **68**, 953–957.

Cain, A. J. (1964). The perfection of animals. *Viewpoints in Biology*, **3**, 36–63.

Clapham, A. R., Tutin, T. G. & Warburg, E. F. (1962). *Flora of the British Isles*, 2nd edn. Cambridge University Press.

Clegg, L. (1978). *The morphology of clonal growth and its relevance to the population dynamics of perennial plants*. Ph.D. thesis, University of Wales.

Darwin, C. (1859). *The Origin of Species*. Harvard Facsimile, 1st edn., 1964.

Dirzo, R. & Harper, J.L. (1982a). Experimental studies on slug-plant interactions. III. Intraspecific variation in the acceptability of plants of *Trifolium repens*. *Journal of Ecology*, **70**, 101–118.

Dirzo, R. & Harper, J. L. (1982b). Experimental studies on slug-plant interactions. IV. The performance of cyanogenic and acyanogenic morphs of *Trifolium repens* in the field. *Journal of Ecology*, **70**, 119–138.

Dolinger, P. M., Ehrlich, P. R., Fitch, W. L. & Breedlove, D. E. (1973). Alkaloid and predation patterns in Colorado lupine populations. *Oecologia* (Berlin), **13**, 191–204.

Ghiselin, M. T. (1974). *The Economy of Nature and Evolution of Sex*. University of California Press, Berkeley.

Gorham, E. (1979). Shoot height, weight and standing crop in relation to density of monospecific plant stands. *Nature* (London), **279**, 148–50.

Gould, S. J. & Lewontin, R. C. (1979). Spandrels of San-Marco and the Panglossian paradigm—a critique of the adaptationist program. *Proceedings of the Royal Society*, B, **205**, (1161), 581–598.

Harper, J. L., Clatworthy, J. N., McNaughton, I. H. & Sagar, G. R. (1961). The evolution and ecology of closely related species living in the same area. *Evolution*, **15**, 209–227.

Haukioja, E. (1980). On the role of plant defences in the fluctuation of herbivore populations. *Oikos*, **35**, 202–213.

Holliday, R. J. & Putwain, P. D. (1977). Evolution of resistance to simazine in *Senecio vulgaris* L. *Weed Research*, **17**, 291–296.

Huffaker, C. B. (1964). Fundamentals of biological weed control. *Biological Control of Insect Pests and Weeds* (Ed. by P. De Bach), pp. 631–649, Chapman and Hall, London.

Jacob, F. (1977). Evolution and tinkering. *Science*, **196**, 1161–1166.

Jones, D. (1962). Selective eating of the acyanogenic form of the plant *Lotus corniculatus* L. by various animals. *Nature* (London), **193**, 1109–1110.

Jones, M. G. (1933). Grassland management and its influence on the sward. *Journal of the Royal Agriculture Society*, **94**, 21–41.

Levins, R. (1966). Strategy of model building in population biology. *American Scientist*, **54**, 421–431.

Onyekwelu, S. S. & Harper, J. L. (1979). Sex ratio and niche differentiation in spinach (*Spinacia oleracea* L.). *Nature* (London), **282**, 609–611.

Pickering, G. (1961). Language: the lost tool of learning in medicine and science. *The Lancet*, 15 July 1961, 115–119.

Putwain, P. D. & Harper, J. L. (1972). Studies in the dynamics of plant populations. V. Mechanisms governing the sex ratio in *Rumex acetosella* L. and *R. acetosa* L. *Journal of Ecology*, **60**, 113–129.

Stebbins, G. L. (1970). Adaptive radiation of reproductive characteristics in angiosperms. I. Pollination mechanisms. *Annual Review of Ecology and Systematics*, **1**, 307–326.

Stebbins, G. L. (1971). Adaptive radiation of reproductive characteristics in angiosperms. II. Seeds and seedlings. *Annual Review of Ecology and Systematics*, **2**, 237–60.

Stern Jr, J. T. (1970). The meaning of 'adaptation' and its relation to the phenomenon of natural selection. *Evolutionary Biology*, **4**, 38–66.

Turkington, R. & Harper, J. L. (1979). The growth, distribution and neighbour relationships of *Trifolium repens* in a permanent pasture. IV. Fine-scale biotic differentiation. *Journal of Ecology*, **67**, 245–254.

Watt, A. S. (1934). The vegetation of the Chiltern Hills with special reference to the beechwoods and their seral relationships. *Journal of Ecology*, **22**, 230–270, 445–507.

Watt, A. S. (1940). Contributions to the ecology of bracken (*Pteridium aquilinum*). I. The rhizome. *New Phytologist*, **39**, 401–422.

Watt, A. S. (1945). Contributions to the ecology of bracken (*Pteridium aquilinum*). III. Frond types and the make up of the population. *New Phytologist*, **44**, 156–178.

Watt, A. S. (1957). The effects of excluding rabbits from grassland B (Mesobrometum) in Breckland. *Journal of Ecology*, **45**, 861–878.

White, J. (1980). Demographic factors in populations of plants. *Demography and Evolution in Plant Populations* (Ed. by O. T. Solbrig), pp. 21–48. Blackwell Scientific Publications, Oxford.

White, J. & Harper, J. L. (1970). Correlated changes in plant size and number in plant populations. *Journal of Ecology*, **58**, 467–485.

Williams, G. C. (1966). *Adaptation and Natural Selection.* Princeton University Press, Princeton, N.J.

Wilson, J. R. (1978). *Plant Relations in Pastures.* C.S.I.R.O., Melbourne, Australia.

THE DEVELOPMENT OF RESILIENCE
IN BURNED VEGETATION

D. WALKER

*Research School of Pacific Studies, Australian National University,
Canberra, A.C.T. 2600, Australia*

SUMMARY

The importance of fire in affecting the composition of many vegetation types (e.g. the Californian chaparral) is well-established. Its significance as an influence in other kinds of apparently non-successional vegetation, particularly those types dominated by long-lived species, is neither obvious nor readily testable experimentally. Published pollen diagrams and charcoal particle counts from sites in the conifer-hardwood forests of north America and from Lake George in south-east Australia provide information about the role of fire in these two regions against a variety of time-scales.

It is likely that, in the early post-glacial period, the spread into the conifer-hardwood forests of some species which later became important there was facilitated by infrequent but intense fires. Their incorporation into the vegetation changed the fire regime, increasing the frequency but decreasing the intensity of the disturbance. Thereafter, diversification of forest types accompanied differences in burning histories.

The area around Lake George has experienced several climatic fluctuations during the last 400 000 years. All the cold periods expelled trees from the area. Those warmer periods which preceded about 120 000 B.P. had wet sclerophyll forests. At first, fire was negligible and the flora was fire-intolerant. In the penultimate interglacial, although fire increased, fire-sensitive trees persisted, but by the beginning of the last interglacial such plants were lost from the area. Their place was taken by fire-tolerant and fire-dependent plants and the amount of burning was much greater than before.

The Australian record demonstrates how fire may be the proximate cause of major plant geographical changes. The American data show how resilience to fire may develop in vegetation. This resilience is fostered by diversity in the flora and this usually diminishes if the recurrence of burning is long-delayed, with the result that the vegetation's capacity to respond to the disturbance when it does happen is reduced. The development of dominance by one or a few species is probably a threat to the persistence of the many vegetation types which have developed in balance with recurrent environmental perturbations such as fire.

INTRODUCTION

In describing the processes which affect the levels of species' populations and their relationships in vegetation, we distinguish between those processes which appear to maintain the *status quo* from those which seem to be causing irreversible change: 'maintenance dynamics' as opposed to 'succession dynamics'. Sometimes it is difficult to know which of these we are witnessing. In order to be sure that a vegetation type is experiencing only population changes related to its continuing maintenance, its composition must be seen to vary in a regular, predictable, even if stochastic, fashion; in particular, there must be no net change in composition over a time period equal to the

0262–7027/82/0300–0027$02.00 © 1982 British Ecological Society

life-span of the longest-lived species. This property is often difficult to investigate because plant life-spans are often long. More usually, descriptions of process are contrived from the notional ordering of stands occurring separately today, into cyclical sequences through imagined time. Bolstered by the observation of one stage succeeding another through relatively short time-intervals, this kind of reconstruction is often the best that can be done. But its uncritical application is fraught with dangers, as has been demonstrated in the cases of bog vegetation and hydroseres which accumulate beneath them at least a partial record of their antecedent conditions (Barber 1981 and references therein; Walker 1970). What is now known about the dimensions and rates of environmental change, particularly those of climate, suggests that succession and migration may be commoner today than persistent maintenance. But this begs the question of whether at least some biotic systems may have evolved maintenance dynamics capable of assuring resilience despite these environmental changes.

This paper is about the way in which vegetation responds to change in its physical environment, as resolved on several time-scales, and particularly how, under some circumstances, a perturbation may be incorporated into maintenance dynamics and, in other circumstances, overcome resilience and induce change. Fire is the environmental factor considered. I use the term *resilience* in Holling's (1973) sense as the capacity of an ecological system to respond to a perturbation by returning to its pre-perturbed state, *stability* being measured by the speed with which it does this. Resilience may be sustained by systems of differing stability, but relatively unstable systems are prone to change as a result of repeated, perhaps dissimilar, disturbances.

THE RESPONSE TO FIRE

Fire is now recognized as a significant ecological factor in many of the world's vegetation types. Its early history is obscure and only recently have continuous records of fine charcoal particles, collected alongside pollen counts, been made at a few Quaternary sites. Fire is, and has long been, a favourite tool in man's exploitation of the landscape but this discussion makes no attempt to distinguish between fires set by people and those caused naturally.

Fire is a *relatively* simple source of stress on plants. Its action is sudden and usually short, although the responses it induces may emerge long after the fire has passed. Its main variations are in intensity, frequency and area, although sometimes only a particular stratum of plants might be directly modified. It is strongly affected by topography and weather.

Fire's potency in affecting vegetation composition is clearly demonstrated, on a time-scale well within human ken, in the chamise chaparral of southern California. In a region dominated by *Adenostema fasciculatum*, a shrub with crowns about 1 m wide and 1·5 m high, the summer-deciduous needle-leaves and shredding bark of which provide highly flammable litter, Hanes (1971) pieced together the history of fire response from a study of sites with known fire histories spanning more than 40 yr. Amongst all the species of this chaparral, two main mechanisms for surviving fire are apparent. The first of these is to sprout from root crowns which are not usually destroyed by fire (e.g. *Prunus ilicifolia, Quercus dumosa*) whilst the second is to invest heavily in soil seed banks (e.g. *Salvia mellifera, Ceanothus crassifolius*). Within a year following a fire, all the species characteristic of the chaparral of the locality are growing above ground. Subsequently individuals and ultimately species are lost, sprouters proving more persistent than most obligately seed-reproducing species. For this reason, and because in the most fire-prone areas fires recur every 2–10 yr, these latter species must seed quickly and copiously following a burn. This is all the more important because most of them do not persist as photosynthesizing plants for more than 20 yr in the absence of fire, whereas many of the sprouted shrubs continue to extend their cover and accumulate biomass beyond this time (Fig. 1). *Adenostema fasciculatum,* which sprouts after fire *and* builds up a large seed reserve in the soil in its early years, exemplifies the sprouters which increasingly dominate the vegetation, producing great quantities of litter as individuals become senescent, a fate

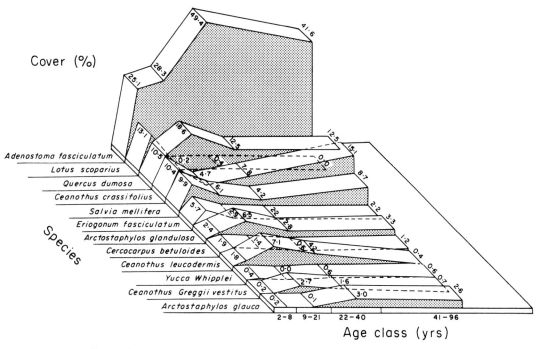

FIG. 1. Changes in cover on coastal exposures of chaparral of the twelve most common species. Percentages are based on mean values in each age class. (From Hanes (1971) by permission. Copyright 1971, the Ecological Society of America.)

which overtakes most of them by about 40 yr after fire. Because of the frequency with which fires recur, this and later stages are rarely reached.

In comparisons of neighbouring stands of chaparral both burned and unburned for 40 yr or more, Christensen & Muller (1975) demonstrated that heat, directly and indirectly, prompts growth initiation in most chaparral plants. After autumn rains, unburned control plots occasionally produced seedlings or vegetative sprouts but these did not survive the following dry season or flower. Fire has several effects which are likely to change this. Hydrophobic substances of plant origin which accumulate near the soil surface are modified and move down the profile increasing the wetability of the uppermost layers. For some species heat breaks the dormancy of their seeds, for others it removes inhibitory chemicals including toxins derived from the chaparral shrubs themselves. Nutrient cycles, blocked by accumulation in the standing biomass and in litter hardly susceptible to direct microbiological action, are temporarily reinstated. The destruction of the litter masses simply vacates surface space.

In the chaparral, fire both devastates the above-ground biomass and promotes subsequent productivity and diversity. There are other controls on the composition of the vegetation (e.g. small mammal grazing (Christensen & Muller 1975); allelopathic leachates from the dominant shrubs (Muller & Chou 1972)) but, without fire, the chaparral as it exists today would not persist. The component species and their reproductive characteristics endow the chaparral with resilience, the mechanism of which is accessible to experimental elucidation because all the species are relatively short-lived and the perturbation is frequent. The existence of similar environmental controls embedded in the dynamics of more complex vegetation with longer-lived plants can only be exposed by palaeoecological techniques which may also hint at the sequence of events by which resilience of the vegetation has developed and the circumstances under which it may be lost. Some insights into these processes are provided by studies from the eastern United States-Canada borderlands and south-eastern Australia.

The conifer-hardwood forests of North America

The conifer-hardwood forests extend in a belt about 500 km wide, roughly centred on the 45°N latitude, from west of the Great Lakes to the east coast of continental America. They occupy land unforested at the height of the last major glaciation, forming a patchy vegetation, the size and composition of units in the mosaic varying substantially in relation to soil parent material, topography and local climate. Less immediately evident determinants of its variability include the timing of post-glacial immigration of its components (Davis 1976) and the effects of fire. In a remarkable investigation of historical records, stand ages and the dendrochronological dating of fire-scars, Heinselman (1973) has conclusively demonstrated the importance to the formation of the pattern of frequency, intensity and extent of fires since the end of the sixteenth century, in one 530 000 ha region of this forest which includes the 215 000 ha reserve called the Boundary Waters Canoe Area of north-eastern Minnesota. Practically all the forest studied was burned at least once between A.D. 1595 and 1972 and the oldest trees presently growing began to do so after the 1595 fire. In the forest mosaic, patches dominated by *Pinus banksiana* with serotinous cones are most common, followed by *Populus tremuloides* and *Betula papyrifera* which sprout from roots or stems after fire, grow quickly and fruit young. Other broad-leaved trees and shrubs (e.g. *Acer rubrum*, *Corylus cornuta*, *Quercus rubra*), which are locally common, have similar responses to fire. *Pinus resinosa* and *P. strobus*, on the other hand, have a resistance strategy, depending on thick bark to mitigate the effects of fires. After a fire the suckering broad-leaved trees usually form a pioneer stand. Within a century these trees have formed a forest with an understorey to which *Abies balsamea*, *Picea* spp. and *Thuja occidentalis* contribute. At this stage birch bark, dead timber from the pioneer trees and flammable conifer foliage begin to provide substantial fuel. This is augmented as the pioneers die and dominance passes to the conifers; total dry biomass reaches a maximum between 200 and 300 yr after fire. Local circumstances impose great variety on this sequence and under appropriate weather conditions all but the earliest stages of post-fire revegetation may carry another fire. This fire-related vegetation has evidently been living in northern Minnesota, and presumably throughout the conifer-hardwood forest range, for at least three centuries, almost the maximum age of the longest-lived trees. Is it an even longer-established, self-maintaining, resilient system?

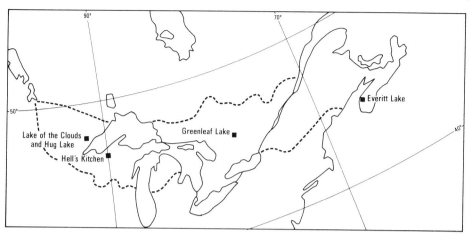

FIG. 2. Location of pollen analytical sites and the approximate boundaries of the conifer-hardwood forests of North America.

Four lakes have provided pollen and charcoal diagrams from which to attempt a reconstruction of the effects of fire on the conifer-hardwood forests (Figs 2 and 3): Lake of the Clouds (Craig 1972; Swain 1973) and Hug Lake (Swain 1980), both in the Boundary Waters Canoe Area of north-eastern Minnesota, Greenleaf Lake (Cwynar 1978) in the Algonquin Park, Ontario, and Everitt Lake (Green 1981, 1982) in south-western Nova Scotia. Hell's Kitchen, from the more strongly hardwood forest of Wisconsin (Swain 1978), also provides some relevant data.

Lake of the Clouds lies in a conifer-rich part of the forest. Its pollen diagrams (Craig 1972) together span a period from about 11 000 B.P. to A.D. 1970 and were dated by counts of annual laminations in the mud. Charcoal is certainly present continuously from about 10 000 B.P. and perhaps earlier (Swain 1973). During the period since then, resolved at 200- to 500-yr intervals, charcoal abundance shows some parallelism with major events

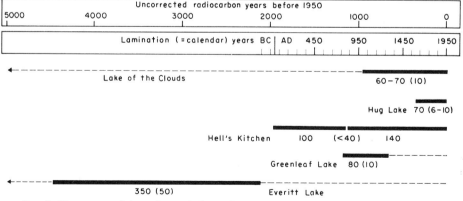

FIG. 3. Time spans of the pollen and charcoal records from the conifer-hardwood forests of North America. For each site, the horizontal bar represents the range of detailed pollen and charcoal analyses, the broken lines the full chronological extensions of the basic pollen diagrams. Numbers below the bars are intervals between fires in years estimated from the charcoal counts; those in parentheses show the mean sampling interval in years. The radiocarbon time scale refers to the long sequence from the Lake of the Clouds and Everitt Lake. The lamination scale refers to Greenleaf Lake, Hell's Kitchen and the youngest part of the Lake of the Clouds.

in the *Pinus* curve but is otherwise difficult to relate to the percentage pollen diagram. Changes in the pollen diagram are attributed mainly to climatic change.

The period from A.D. 980 to 1970, however, is more finely resolved (Swain 1973), intervals between samples ranging from 10 to 30 yr, and both charcoal and pollen are expressed as influx values (grains per unit area of sediment surface per unit time). The smoothed charcoal curve falls at about A.D. 1500 to values about 70% of earlier levels, which may have some regional significance, but the actual curve is irregular with some well-defined peaks, some formed by more than one sample (Fig. 4). Peaks of charcoal closely followed by increases in lamina thickness, indicative of increased erosional input to the lake after fire, are taken to represent fires in the neighbourhood of the lake. Several of the more recent of these correlate well with historically or dendrologically authenticated fires but some of the latter would not have been identified from the charcoal curve, inviting

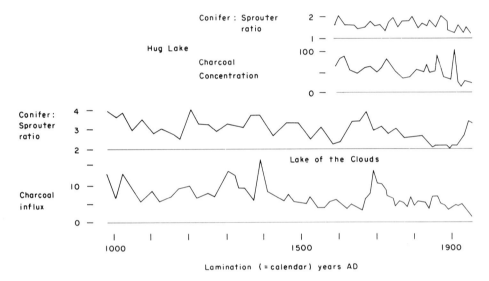

FIG. 4. Relationships between charcoal and tree regeneration index (conifer : sprouter ratio) at two sites in the Boundary Waters Canoe Area of Minnesota. (Redrawn from Swain 1973, 1980.) Charcoal peaks, particularly those associated with unusually thick laminae at the Lake of the Clouds, are followed by rises in the conifer : sprouter ratio at Hug Lake and falls in the ratio at the Lake of the Clouds.

the supposition that fires have been more frequent than the charcoal peaks imply throughout the last millennium. Nevertheless, estimates of the total number of fires in this period range only between 14 and 17 and the corresponding mean fire frequencies between 1 in 71 and 1 in 59 yr; intervals between successive fires vary from a decade to a century or more. The responses of individual taxa to separate fire events are almost impossible to derive from the data as presented. If, however, the deciduous trees, bracken and grasses, which are capable of vigorous vegetative regrowth ('sprouting') and flowering after a burn, are summed, they often attain ascendancy over the conifers, which reproduce only from seed, for 20–30 years (Fig. 4).

Hug Lake is about 22 km south-east from Lake of the Clouds, in gentler topography. Apart from a few groves of *Abies balsamea*, *Picea glauca* and *P. mariana*, a mixed deciduous facies of the forest extends for 2·4–6·4 km from the lake depending on direction; logging has not approached closer than 2 km to the lake, and that only since

1940. The pollen diagram is dated by an ingenious but not entirely dependable method, so that neither charcoal nor pollen influx can be reliably calculated (Swain 1980). Nevertheless, changes in charcoal *concentration* in the sediment and a conifer:'sprouter' pollen ratio can be compared confidently. The diagram, which runs from an estimated A.D. 1600 to A.D. 1970 sampled in decades, is dominated by far-travelled, regional, conifer pollen. The response of the conifer:'sprouter' ratio to peaks of charcoal is consistent throughout the period but the opposite of that experienced at Lake of the Clouds (Fig. 4). This is interpreted to indicate that conifers had been remote from Hug Lake at least since A.D. 1600 and that fires in the lake's vicinity had only deciduous 'sprouters' to incapacitate. The effect was to reduce their pollen output for about 20 yr whilst the regional conifer pollen input remained unaffected, resulting in a higher conifer:'sprouter' ratio after fires. Because the fire frequency has been about the same at the two sites, it is suggested that topography and related factors might be the prime determinant of the vegetation differences around Hug Lake and Lake of the Clouds. Yet, protected from fire, as at present, spruce and fir can reach the Hug Lake forests, for seedlings of these trees have been found there. The reasons for the exclusion of conifers from the Hug Lake area are obviously complex; nevertheless it is extremely interesting that the different balances between major taxa around the two lakes have developed and persisted under broadly the same fire frequencies. In the two groups of trees, selection in response to fire has led to two quite distinct survival strategies. Not only do these seem to be equally effective when their owners grow together but one, the 'sprouter' strategy, can operate successfully alone. Perhaps with this fire regime, 'sprouting' hardwood or mixed forests are possible whereas purer conifer forests are not.

Greenleaf Lake, Ontario, lies in a complex patchwork of conifer-hardwood forest in which pines (*Pinus resinosa*, *P. strobus*) are particularly abundant. The pollen diagram (Cwynar 1978), beginning *c.* A.D. 770, suggests that the vegetation has not changed markedly and permanently since then, despite the fact that fire frequency between A.D. 770 and 1270, as estimated from charcoal influx, seems to have been 1 in *c.* 80 yr whereas the incidence based on fire-scars for the most recent 300 yr is 1 in 45 yr. Either the charcoal-based frequency is an underestimate or the forest's maintenance mechanism is resilient to, and apparently equally stable under, fire frequencies differing by a factor of two. In the period analysed at decadal intervals, between A.D. 770 and 1270, charcoal peaks indicate six major fires. The *Pinus*:*Betula* pollen ratio, which Cwynar judges most likely to indicate vegetation response to fire, falls after only three of these peaks. Indeed, after one of them the ratio rises to unprecedented heights before falling steeply coincident with the following charcoal peak. If this event is dismissed as being due to some unknown and infrequent perturbation to which the vegetation is less stable, the normal response mechanism to fire must operate within the sampling interval of 10 yr; in this respect the vegetation is very stable. Alternatively, as Cwynar suggests, local vegetational effects comparable with charcoal influx may be masked by more regional features of the pollen rain.

Hell's Kitchen, Wisconsin, is set in a forest with only about 15% of conifers. In its pollen diagram Swain (1978) has identified a sequence of events following each fire from about 65 B.C. to A.D. 1973, eighteen in all. Although charcoal was present throughout the sediments, changes in its quantities in the analyses did not allow it to be used as a fire marker. Instead, peaks in *Betula* and *Populus*, which respond quickly after burning, were taken to mark fires. Observations of the living forest indicate that *Pinus strobus* and *Tsuga canadensis* will replace *Betula papyrifera* in the absence of fire and, indeed, in all but a very few cases, they are seen to do so in that order throughout the pollen diagram. On average, *Pinus* maxima occur 40–80 yr after those of *Betula* and 80–170 yr before the succeeding *Tsuga* peak (Fig. 5). In the climatically drier interval before about A.D. 825 (i.e. c. 1150 yr before 1973), the average fire frequency was 1 in 100 yr and some fires were evidently close enough in time to prevent *Tsuga*, and in one case *Pinus*, recovering. Subsequently, fires occurred less frequently (1 in 140 yr) in a moister climate and the sequence to *Pinus* and *Tsuga* was always completed but the fire-dependent *Populus* was less consistently present. This suggests that an established assemblage of species can remain resilient, not only under relatively small and slow climatic fluctuations but also after a changed periodicity in fire impact; in the present case this is assisted by the differing ecologies of *Betula lutea* and *B. papyrifera*. Only a small proportion of the total changes in the regeneration system can be inferred from the pollen diagram but these seem to have included reductions in the abundance of *Populus* and a roughly 25% loss of stability as measured by the time taken for the fire-induced *Betula* pollen peak to be followed by that of *Tsuga*, the greatest component of this being in the *Betula* to *Pinus* interval.

Everitt Lake, from the conifer-hardwood forests of Nova Scotia, has provided a pollen diagram running from about 11 000 B.P. to the present (Green 1981). Using time-series

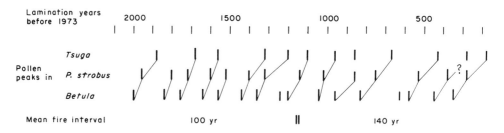

Fig. 5. Sequences of related peaks in pollen curves probably resulting from repeated fires at Hell's Kitchen, Wisconsin. (Drawn from data in Swain 1978.)

analysis and other statistical techniques (Walker & Wilson 1978; Walker & Pittelkow 1981) on these data, Green (1981, 1982) has obtained astonishingly suggestive indications of the relationships between forest trees and fire. Separate analysis of the pollen abundances of each taxon throughout the core not only identified periods of major change in the vegetation, the most general being at about 9200 and 3900 B.P., but also established the positions of statistically significant changes in abundance of each individual taxon; these did not necessarily coincide between taxa (Fig. 6). Comparison with peaks in the charcoal curve then demonstrated that the actual arrival of some important trees at the site (e.g. *Abies, Acer, Fagus, Fraxinus, Ostrya-or-Carpinus, Tsuga, Ulmus*) followed major fires. Some of these (e.g. *Fagus, Tsuga*) were then able to survive at very low abundances for many centuries before another major fire, under climatic conditions that now favoured them, gave them advantage over their erstwhile more successful competitors. These correlations with fire are strongest from 11 000 to 9000 B.P., remain strong but lag a century behind from about 9000 to 4000 B.P., and after about 4000 B.P., when there are no large charcoal peaks, are entirely absent. The explanation must be that vegetation changes of this magnitude and persistence are triggered only by 'big' fire events which also overcome the inertia imparted by the capacity of large, long-lived, established species to withstand the slow deterioration of their environment (e.g. by small but maintained climatic change) and prevent younger, smaller but otherwise better fitted, plants from gaining abundance. Green (1982) suggests that before 6000 B.P. the climate of Nova Scotia and the fuel produced by the dominant conifers combined to allow frequent *large* fires, as evidenced by the numerous charcoal peaks (Fig. 6). Major perturbations of this kind, with a mean frequency of about 1 in 300 yr, 'opened' the existing system to newcomers which were migrating into the region following the last deglaciation. Practically every major fire episode therefore removed any developing resilience since each new assemblage, as it grew up, was unstable to another fire because its composition was predicated by different prior conditions. However, after 6000 B.P. no more species arrived, there was less conifer fuel and probably more stable weather patterns. Only three major fires comparable with the earlier ones occurred in the 6000-yr period to the present day, an order-of-magnitude fall in frequency. There was therefore every opportunity for the vegetation to adapt to the very frequent occurrence of local, smaller, less intense fires

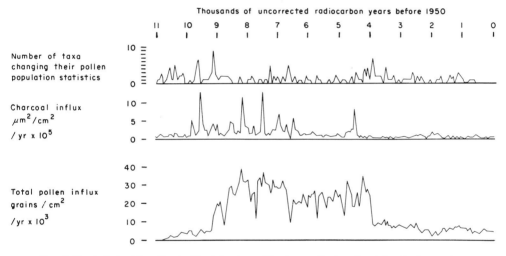

FIG. 6. Everitt Lake, Nova Scotia. The behaviour of fourteen pollen taxa in relation to changes in charcoal influx. Significant changes in the population statistics of each separate taxon were determined and the number of taxa exhibiting such changes at each level (date) plotted on the top graph. (Redrawn from Green 1982.)

during that period, drawing on the variety of responses of which the now consistently large flora was capable.

More detailed analysis by power spectrum and cross-correlation techniques enabled Green (1981) to expose the mechanism by which resilience was maintained during the period 4450–2100 B.P. after the major change of behavioural mode of the vegetation. Charcoal abundance usually recurs at about 350-yr intervals, though sometimes at 165-yr intervals. Total pollen influx to the sediment has a 210-yr recurrence period but correlates strongly and negatively with charcoal abundance at zero and 350 yr behind the latter, thereby showing immediate response of pollen to charcoal. The greatest influxes of most separate pollen taxa occur with the same frequency as do high charcoal levels but also at other frequencies suggesting some complexity in their control. Cross-correlograms between charcoal and the abundance of pollen of individual taxa (Fig. 7) analyse this reponse and in particular specify the most common time-lag between the incidence of fire and the abundance surge of each taxon. Of the plants illustrated, none has its maximum positive correlation with charcoal at zero lag, but *Acer*, *Betula* and *Fraxinus* have significant negative correlations there, indicating that, within the limits of sampling interval (50 yr), these three are devastated at the time of fires. Others (e.g. *Picea, Pinus, Ulmus*) have positive correlations with charcoal at 50 yr lag, others (e.g. *Fraxinus, Quercus*) at 150–200 yr and yet others (e.g. *Abies, Acer, Fagus*) after 300–350 yr when the next fire is most likely to be imminent. Cross-correlograms between taxa, without reference to charcoal, similarly sort out the relationships of their temporal behaviour and might even suggest spatial associations, particularly between taxa which also have the same cycling period. The capacity of this technique to draw from a pollen diagram the preferred sequencing of fluctuations in plant taxa and the relationship of this to fire events, all on a known time-scale covering more than 2000 yr, permits the modelling of vegetational events

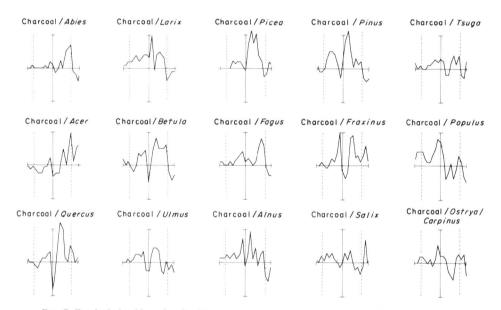

FIG. 7. Everitt Lake, Nova Scotia. Cross-correlograms between charcoal and a range of pollen taxa. On each diagram the vertical scale measures correlation and the horizontal scale time before and after charcoal maximum marked by the continuous vertical line (dashed line marks 350-yr interval). (Redrawn from Green 1982.)

following fire with some confidence (Fig. 8). It is, of course, a complex of possibilities to which transition probabilities cannot be allocated from the present data. But the variety of courses between defined stages, deriving from the number of taxa, their differing longevities and the variety of their responses to their physical and biotic environments, ensures that the system is resilient, at least for so long as the interval between fires of this 'lesser' magnitude does not greatly exceed 350 yr.

The data from the four main sites and Hell's Kitchen are of varied quality, obtained by differing procedures and do not all overlap in time. Importantly, different fire sizes and frequencies are clearly involved. At Everitt Lake very 'big' fires occurred at 300-yr intervals before 6000 B.P. and at 3000-yr intervals thereafter; the distinction cannot be made in the only other comparably long record from Lake of the Clouds. Also at Everitt Lake, a frequency of one smaller fire peak in 350 yr was the rule before about 2000 B.P., after which the data suggest the possibility of a greater fire-frequency which might have gone with a switch from a fairly uniform vegetation to the complex mosaic of more recent times. The other sites, details from all of which belong after 2000 B.P., might well reflect circumstances after such a switch with frequent, if still less devastating, fires at frequencies between 1 in 70 and 1 in 140 yr. In the earlier period to about 6000 B.P., the dominant trees produced a lot of fuel, formed a closed canopy and were potentially long-lived. There is no knowing whether they would have formed a persistent, self-maintaining, vegetation although, without competition from new additions, there was no climatic reason why this should not have happened. But newcomers were constantly arriving and big fires every 300 yr allowed them to establish and compete with the former dominants. Under these circumstances the assemblage was constantly changing; there was no resilience or stability but the perturbations were crucial to the speed with which a vegetation of more enduring composition was synthesized. Once this had happened, with the resulting slower build-up of fuel, fires remained about as frequent but were less intense or widespread or destructive than before. The larger number of species and their diverse responses to fire impact

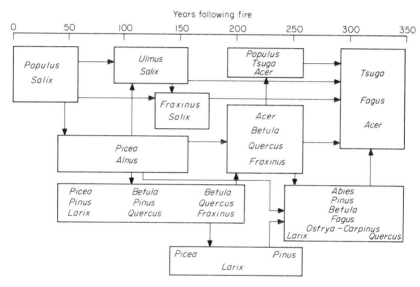

FIG. 8. Temporal relationships of tree genera following fire, derived from the Everitt Lake pollen diagram for the period 4450 to 2100 B.P.. (Redrawn from Green 1981.)

provided much greater variety of courses for secondary successions to take following a fire event. The processes were there in sufficient variety to give maintenance dynamics a stochastic quality. Under these circumstances stability would vary greatly according to the particularities of an individual fire but resilience was evidently high. Indeed, the way in which fluctuations in the populations of critical trees can be related statistically suggests that the forests of the time were rather uniformly mixed. In the rare event of a big fire occurring in such a forest, as at 4500 B.P. at Everitt Lake, major changes were recorded in the behaviour of a large number of tree pollen taxa. In many of these cases, however, the initial change was rapidly followed by another one and, within a millennium, balance was re-established. This implies resilience of a very high order achieved through varying rates of recovery following different levels of a perturbation.

Changes in the statistical associations between pollen taxa at Everitt Lake suggest that at about 2000 B.P. the vegetation might have became more grossly patchy, comparable with the patterning throughout the conifer-hardwood forests at the present day. If this were the case, the accumulation of fuel would be equally patchy, leading to local differences in the incidence and intensity of fires. The dominant frequencies of 1 in 70 to 1 in 140 yr in parts of this period at the Lake of the Clouds, Hug Lake, Greenleaf Lake and Hell's Kitchen are 2–4 times greater than immediately before at Everitt Lake. This change may be related to a coarsening of the vegetation mosaic; it may also be an artefact of the wider sampling interval at Everitt Lake. One may speculate, however, that topographic detail might predispose some sites to more frequent ignition than others, more often curtailing the succeeding regeneration sere. The maximum period between fires at any place in the vegetation would not change, so that the mean frequency overall would increase. Localization of frequency of perturbation would lead to concentrations of particular species groups and to the coarser mosaic of the present day. Such differentiation of the vegetation carries with it the risk of loss of resilience if it reaches a point where seed- or vegetative-parents of all the trees in the system are not sufficiently close to each patch after a burn. Patches to which slowly-responding species did not have access would not be any less resilient, provided that they continued to sustain frequent fires; in a sense such a situation seems already to have been achieved near Hug Lake by A.D. 1600. But patches which lost rapidly-regenerating species would be forced back toward the big fire regime of the earlier part of post-glacial time. It is worth remembering the loss of stability associated with decreased fire frequency at Hell's Kitchen and Heinselman's (1973) observation that protection of the living forests from fire will lead to a loss of species, increase the probability of occasional but big fires and block nutrient cycling by the long accumulation of litter.

Seen in this long historical perspective, it is evident that fire first facilitated the synthesis of the conifer-hardwood forests but did so selectively because only plants capable of withstanding big and frequent fires could persist. The accumulation of such species, and perhaps greater climatic stability, led, after five millennia, to a diverse but probably rather fine-grained forest, carrying equally frequent but less devastating fires to which it was strongly resilient by virtue of the number of species and also the variety of their responses to fire and their ready access to any burned site. For unknown reasons the spatial distribution of taxa may have become more coarsely patchy during the last 2000 yr with increased mean fire frequency but increased threat to overall resilience.

Lake George, south-east Australia

Lake George is situated near Canberra on the southern tablelands of south-east Australia and receives about 600 mm of rain a year. The pre-European vegetation of its surroundings was a mosaic of open grassy woodlands and low open forest. The closest tall wet sclerophyll forest is now 50–70 km to the south-east where rainfall exceeds 700 mm, or more commonly 800 mm, a year. Singh (Singh, Kershaw & Clark 1981) has identified three interglacial periods preceeding the last glaciation and its interstadial stages in a pollen diagram which, with considerable justification, is thought to span about 400 000 yr (Fig. 9). Before about 120 000 yr ago, the warmer interglacials and interstadials were marked by fire-intolerant plants, particularly *Casuarina*, *Cyathea*, *Drimys* and *Podocarpus*, perhaps best interpreted as belonging to tall wet forest. The colder periods, on the other hand, seem to have been treeless but with some *Cyathea* and shrubs. The beginning of the last interglacial heralded a changed system; *Eucalyptus* and other Myrtaceae became important and *Acacia* appeared, whilst *Casuarina*, *Cyathea*, *Drimys* and *Podocarpus* had much diminished roles. Cold periods, as before, were treeless. Finally,

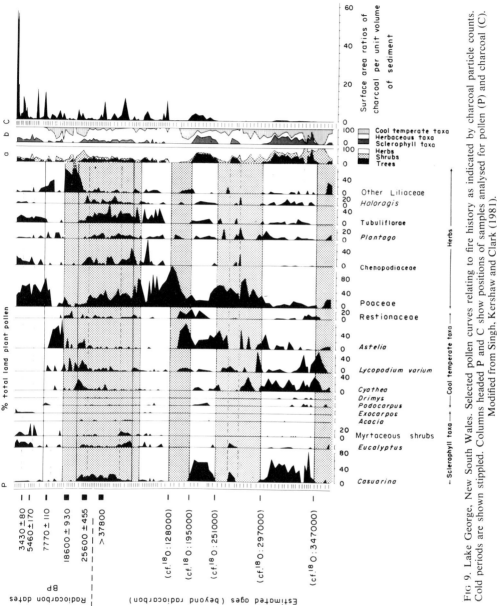

FIG 9. Lake George, New South Wales. Selected pollen curves relating to fire history as indicated by charcoal particle counts. Cold periods are shown stippled. Columns headed P and C show positions of samples analysed for pollen (P) and charcoal (C). Modified from Singh, Kershaw and Clark (1981).

during the last 8000 yr, there have been no cool-temperate, fire-intolerant plants in the Lake George pollen catchment.

The abundance of charcoal particles in the samples counted for pollen is strikingly distributed. Before about 120 000 yr ago it was intermittent and low; after that time it was virtually continuous, fluctuating greatly but reaching unprecedentedly high levels during the last 200 yr since European settlement. In the earlier period such charcoal as was recorded came from the interglacial periods when there was forest to be burned, albeit grudgingly one imagines. In the penultimate interglacial there is more charcoal implying a greater impact of fire, but fire-intolerant plants persisted through its 50 000 or more years. Subsequently, however, and particularly from the beginning of the last interglacial, the continuous occurrence of charcoal is strikingly parallelled by the absence of fire-intolerant plants, and, in the last 6000 yr or so, by high values of Myrtaceae and *Acacia*.

It cannot be determined whether more frequent or intense fires were the prime agent of change in the region or whether some other, perhaps climatic, factor altered, bringing fire-tolerant vegetation from the drier west and with it more fire. If the former were the case, then some persistent fire-promoting agents, such as men or, again, climatic change, must have been at large. But it is evident that, whether beset by other disadvantages or not, vegetation which had recurred under suitable climatic conditions for at least 200 000 years was displaced and that fire was one of the agents active in its regional extermination. Whereas the vegetation could tolerate some small amount of fire by a response mechanism which is beyond resolution, its resilience was completely overcome by the diabolical conditions which supervened. *Casuarina* persisted, perhaps along watercourses, in reduced quantity until about 20 000 years ago, when it too succumbed. The Lake George record is, in a sense, crude but it illustrates how, viewed over hundreds of thousands of years, fire has exercised very strong selection pressure, replacing vegetation whose maintenance dynamics were stable under slight stress from fire by vegetation which is strongly adapted to it. For there can be little doubt that, protected from fire, the vegetation around Lake George today would change to something in which fire-adaptations such as lignotubers, epicormic shooting, thick insulating bark, serotinous seeds and many life-cycle phenomena (Gill 1981) conferred no special advantage.

CONCLUSIONS

The recurrent oscillations of major vegetation formations which dominate the Lake George pollen diagram of the last 400 000 yr are direct evidence of vegetation changes which can be parallelled in many parts of the world. Each major oscillation is both similar to and different from that which preceded it. Cumulative differences which appear over hundreds of thousands of years are the results of migration under persistent, if fluctuating, environmental stresses such as climatic change. Sometimes these stresses may act directly but, more often, operate on plant assemblages through other agencies. Whether it was indeed a general change in the nature of interglacial climates or the purposive activities of human beings which introduced fire to the Lake George region, fire was the direct cause of vegetation change. Its recurrent impact not only replaced a fire-sensitive vegetation by a fire-tolerant one but also effected subsequent changes in the latter as burning itself altered in frequency and intensity. The Lake George record also demonstrates the importance of an appropriately adapted flora to take advantage of new circumstances without itself being destroyed by them. In this case the flora was the product of some 10 million yr of selection for tolerance to intense seasonality, drought and heat and, presumably, fires.

The actual processes through which these balances and imbalances are attained is more sharply demonstrated in the records from the conifer-hardwood forests. Initially, the perturbation itself creates opportunities for new immigrants by repeatedly destroying stands of the existing vegetation. Of the flora available to colonize these gaps, only species which can tolerate the perturbation will persist. In time, the destructive effects of the perturbation are ameliorated by the accumulation of more species with diverse responses to fire-associated events. At Everitt Lake this evidently happened in the first 5000 yr of post-glacial vegetation. This diversity seems to be essential to the maintenance of resilience in systems of this kind. It may be expressed through more or less evenly mixed stands or in a mosaic of patches each with different fire history and therefore different flora and future burning potential.

If a perturbation fails to recur for an unusually long time, diversity is reduced and a few species assert their dominance. This is equally the case, on time scales reckoned in plant generation-times, in the conifer-hardwood forests of the past with a dominant fire interval of 350 yr, in particular areas of that forest today with but 70 yr between major fires and in the chaparral which burns much more frequently. It is also the case on raised peat bogs, where clear stratigraphic evidence shows that increased water input flooding hollows and stimulating hummock growth, through increasing environmental diversity accommodates more species than can either the prior or subsequent conditions. Because of the variety of responses it offers, diversity can tolerate variation in the incidence, intensity and size of the perturbation much more effectively than can a small number of species or strongly determinate response-paths. To a great degree it is the success of the diverse phase in a regeneration sequence which allows a force which stressed a former vegetation to be accommodated; it is the essence of resilience.

There can be no doubt that all of these fire-tolerant vegetation types could be disrupted by unusually frequent or intense burning. An equal threat to their resilience, however, seems to come from unusual delay in the recurrence of the perturbation. The diverse phase is followed by loss of species through competition and the emergence of a few dominants. The longer this tendency persists without a recurrence of perturbation, the less likely are the early stages to regenerate. A fire in a 300-yr-old unburned patch of conifer-hardwood, dominantly *Picea* and *Abies* producing masses of flammable litter, is likely to be destructively intense, whilst the stocks of pioneer trees will have fallen through the passage of time and the long domination of a few conifer species. Chaparral which has escaped fire for more than about 60 years is composed almost exclusively of senile *A. fasciculatum* bushes presiding over much depleted soil banks of species which have not photosynthesized in that spot for half a century. A burn will revivify the *A. fasciculatum*, but cannot replace species of which even the seeds have died. A bog spared surface water for an unusually long interval and covered with *Calluna vulgaris*, *Eriophorum vaginatum* and little else, will have no way of responding biotically to a rise in rainfall; it will erode.

The emergence of dominance is therefore a serious threat to the resilience of vegetation systems attuned to periodic perturbation. It may of course be partially controlled by other factors: disease in conifers, fire on bog surfaces. Its effects are ameliorated if they are expressed patchily.

The capacity to pre-empt space and environmental components such as nutrients so as to exclude other plants from localities may also, under some circumstances, conserve resilience in a changing environment, particularly when the species involved are gregarious and can regenerate beneath their own parents. When such conservative systems finally succumb to environmental stress, however, they are often replaced by others of greater diversity. When there is a steady, persistent drift in an environmental variable such as temperature with a direct effect on plants, single-dominant vegetation types might be common; but they are successional, they lack resilience. Pollen diagrams from early post-glacial times throughout the world abundantly demonstrate this.

It may be that dominance is primarily an adaptation to strong environmental stresses of particular kinds whilst diversity is favoured by apparently more stable circumstances which, however, contain recurrent perturbations. Comparison of appropriate sections of Quaternary pollen diagrams with supposed analogous vegetation types at the present day certainly suggests this. From the example of the present 'interglacial', the superficially simpler and certainly longer-lasting vegetation types filling the middle periods of interglacials seem to be the more complex dynamically than those at either end. But both these extremes are adapted to environmental changes occurring with frequencies and amplitudes which are unusual in the whole history of terrestrial vegetation.

In the remote past, vegetation was probably subject to recurrent small perturbations and to long-maintained but slow environmental drifts. During the later part of the Tertiary the environmental drifts became faster and, with more intense changes in the physical landscapes of the time, major disruptions to the vegetation occurred more commonly. In the Quaternary, rapid oscillations have replaced uniform trends in environmental change and some of these have enormously increased the frequency of vegetation-devastating events and the creation of new habitats. These are expressed on many scales, from a fire through a forest to the glaciation of a continent. All leave gaps to be filled. All give selective advantage to opportunists. For the hiatus caused by a major stress is no sooner filled than another perturbation, perhaps on a different scale, follows in its wake or an apparently non-oscillatory trend begins to change direction. Opportunism combined with mechanisms to resist extinction between perturbations, characteristics most often associated with ruderals and 'weeds', is at a premium in the Quaternary. It even dominates the maintenance dynamics of our apparently more resilient vegetation types.

ACKNOWLEDGMENTS

I am grateful to Dr D. G. Green for access to his unpublished work; to him, Dr G. Singh and Dr E. I. Newman for comments on the manuscript; to T. Baumann who drew the diagrams and to Caroline Twang who typed the text.

REFERENCES

Barber, K. E. (1981). *Peat Stratigraphy and Climatic Change*. A. A. Balkema, Rotterdam.
Christensen, N. L. & Muller, C. H. (1975). Effects of fire on factors controlling plant growth in *Adenostoma* chaparral. *Ecological Monographs*, **45**, 29–55.
Craig, A. J. (1972). Pollen influx to laminated sediments: a pollen diagram from northeastern Minnesota. *Ecology*, **53**, 46–57.
Cwynar, L. C. (1978). Recent history of fire and vegetation from laminated sediment of Greenleaf Lake, Algonquin Park, Ontario. *Canadian Journal of Botany*, **56**, 10–21.
Davis, M. B. (1976). Pleistocene biogeography of temperate deciduous forests. *Geoscience and Man*, **13**, 13–26.
Gill, A. M. (1981). Adaptive responses of Australian vascular plant species to fires. *Fire and the Australian Biota* (Ed. by A. M. Gill, R. H. Groves & I. R. Noble), pp. 243–271. Australian Academy of Science, Canberra.
Green, D. G. (1981). Time series and postglacial forest ecology. *Quaternary Research*, **15**, 265–277.

Green, D. G. (1982). Pollen time series and postglacial forest dynamics. *Journal of Biogeography*, (in press).

Hanes, T. L. (1971). Succession after fire in the chaparral of southern California. *Ecological Monographs*, **41**, 27–52.

Heinselman, M. L. (1973). Fire in the virgin forests of the Boundary Waters Canoe Area, Minnesota. *Quaternary Research*, **3**, 329–382.

Holling, C. S. (1973). Resilience and stability of ecological systems. *Annual Review of Ecology and Systematics*, **4**, 1–23.

Muller, C. H. & Chou, C.-H. (1972). Phytotoxins: an ecological phase of phytochemistry. *Phytochemical Ecology* (Ed. by J. B. Harborne), pp. 201–216. Academic Press, London.

Singh, G., Kershaw, A. P. & Clark, R. (1981). Quaternary vegetation and fire history in Australia. *Fire and the Australian Biota* (Ed. by A. M. Gill, R. H. Groves & I. R. Noble), pp. 23–54. Australian Academy of Science, Canberra.

Swain, A. M. (1973). A history of fire and vegetation in northeastern Minnesota as recorded in lake sediments. *Quaternary Research*, **3**, 383–396.

Swain, A. M. (1978). Environmental changes during the past 2000 years in north-central Wisconsin: analysis of pollen, charcoal, and seeds from varved lake sediments. *Quaternary Research*, **10**, 55–68.

Swain, A. M. (1980). Landscape patterns and forest history in the Boundary Waters Canoe Area, Minnesota: a pollen study from Hug Lake. *Ecology*, **61**, 747–754.

Walker, D. (1970). Direction and rate in some British post-glacial hydroseres. *Studies in the Vegetational History of the British Isles* (Ed. by D. Walker & R. G. West), pp. 117–139. University Press, Cambridge.

Walker, D. & Pittelkow, Y. (1981). Some applications of the independent treatment of taxa in pollen analysis. *Journal of Biogeography*, **8**, 37–51.

Walker, D. & Wilson, S. R. (1978). A statistical alternative to the zoning of pollen diagrams. *Journal of Biogeography*, **5**, 1–21.

E. I. Newman (ed.) *The Plant Community as a Working Mechanism*

ON PATTERN AND PROCESS IN FORESTS

T. C. WHITMORE

Commonwealth Forestry Institute, University of Oxford, Oxford OX1 3RB, U.K.

SUMMARY

It is suggested that forests throughout the world are fundamentally similar in their patterns in space and time because the same processes of succession and maintenance operate. The patterns and processes of change in a forest are expressed by its growth cycle in which three arbitrary phases can be distinguished: gap, building and mature. Gaps vary from a few metres to several kilometres across. Examples are given of gap creation at all scales in north temperate, Chilean and tropical rain forests.

In nearly all forests there are different ecological groups ('tolerance classes') of tree species adapted to regenerate in gaps of different size. The existence of these groups was realized long ago by foresters. The principal characteristic of each group is the amount of light required by seedlings. Others concern seed biology, tree architecture and aspects of ecophysiology.

In secondary succession dominance is by progressively more shade-tolerant species. The stage which is richest in species depends on the numbers of species belonging to different tolerance classes. In very species-rich tropical rain forests many species may be adapted to a gap of any given size, i.e. have overlapping niches.

Man has many uses for forests. The yield of different products is maximized by variously manipulating the forest growth cycle. Conservation necessitates maintenance of the full species-richness which is dependent on the continuing occurrence of gaps of all sizes. Repeated, frequent disturbance, especially that which creates large gaps, can lead to floristic impoverishment.

INTRODUCTION

In this paper I suggest that forests throughout the world are fundamentally similar, despite great differences in structural complexity and floristic richness, because processes of forest succession and many of the autecological properties of tree species, worked out long ago in the north temperate region, are cosmopolitan. There is a basic similarity of patterns in space and time because the same processes are at work. My focus is on the dynamics of the structure of the forest canopy and the associated variation in floristic composition. Other aspects of forest dynamics, expressed as phenology, productivity, nutrient cycling, phytophagy and pathogen-attack are not treated. The examples chosen are illustrative only. Mention is made of the studies of A. S. Watt (1919–34) on the semi-natural forest patches ('woodlands') of southern England. These contributed to his later powerful and seminal synthesis, widely referred to as 'pattern and process' (Watt 1947), which is the nub of the present analysis.

In brief, this is an attempt to show how pattern and process, in the sense of Watt, is a concept which gives unity to the range of dynamic processes now described for forests from many parts of the world.

0262-7027/82/0300–0045$02.00 © 1982 British Ecological Society

THE CONTRIBUTION OF FORESTERS

Foresters, as empirical ecologists, have known about the contrasting behaviours of different tree species for as long as they have practised silviculture, that is manipulated forests to encourage some species rather than others. This long predates scientific interest in vegetation dynamics. A selection system was being practised in France in the thirteenth century. Foresters had a well-developed body of practical knowledge by the time ecology as a distinct scientific discipline began to emerge after the invention of the term 'ecology' in the mid-nineteenth century (Haeckal 1869). The idea of forest succession was implicit in the writings of foresters from the eighteenth century onwards (Spurr 1952) and the term itself was apparently introduced by Thoreau in 1863 in a discussion on the forests of New England. In the tropics silviculture became organized in India in the mid-nineteenth century. But foresters are primarily concerned with empirical findings of practical application not with the broad biological implications of their discoveries, and their knowledge of forest dynamics did not for a long time become widely known or discussed amongst botanists.

THE FOREST GROWTH CYCLE

When in a closed forest a tree dies or falls a gap forms. A new tree or group of trees grows up to fill the gap: this is gap phase replacement (Watt 1947). It is convenient to recognize a forest growth cycle (Cousens 1974; Whitmore 1975, 1978) of gap, building and mature phases (cf. Watt 1947). Each phase has a different structure and one develops into the next. Thus the canopy of a forest is continually changing as trees grow up and die.

The environment above and below ground differs between big and small gaps. Conditions in big gaps are more like those outside the forest, in small gaps more like the forest interior. Amongst tree species two extreme states can be seen: those which are adapted to regenerate in open sites and big gaps and those which are adapted to closed forest and small gaps. Response to light is one of the most important attributes of these species groups and various names have been applied to them. The big-gap species are often called light-demanders, intolerant (of shade) species, pioneers or sometimes nomads (van Steenis 1958). The other group are known as shade-bearers, tolerant species, climax species or dryads.

An important characteristic of extreme light-demanding tree species is that they do not perpetuate themselves *in situ* but need open sites, such as large canopy gaps. By contrast seedlings of shade-bearers can establish under shade and have the ability to survive there, awaiting the creation of a gap above them into which they can grow.

A forest consists of a mosaic of structural phases which are always altering as one phase changes to the next. The coarseness of the mosaic depends on what creates the gaps. At one extreme an individual tree dies, at the other a catastrophe such as a hurricane or fire destroys a substantial tract. The floristic composition depends on gap size, with light-demanding species occurring where gaps are large and the structural mosaic is coarse, and shade-bearers where gaps are small and the mosaic fine. Floristic fluctuation occurs where small gaps alternate in time with large ones. Thus, gap-forming processes drive the forest growth cycle and determine forest floristics. To the extent that gaps develop unpredictably there is an element of randomness in forest composition.

There has been considerable discussion as to whether the whole of such a dynamic system should be considered climax forest, or whether climax forest should exclude those parts recovering from catastrophic disruption, which are considered to be undergoing floristic succession not simply cyclical change by gap phase replacement.

It should be noted that any individual tree colonizing a gap contributes as it grows to building and then to mature forest. It is therefore a mistake to refer to gap-phase and mature-phase tree species (*pace* UNESCO/UNEP/FAO 1978, Chapter 8), for this mixes the concept of the forest growth cycle with that of species' autecology.

PATTERN AND PROCESS IN SOME TEMPERATE FORESTS

Beechwoods of southern England

The forest growth cycle and cyclical replacement of different tree species, as exemplified by the beechwoods (*Fagus sylvatica**) of the Chiltern Hills and South Downs of southern England, was one of the seven plant communities with which Watt illustrated his analysis of pattern and process in the plant community (Watt 1947).

Watt described structural changes in beechwoods as the trees grow and correlated changes in the associated ground vegetation. He also recorded places where ash (*Fraxinus excelsior*) and birch (*Betula* spp.) occur in cyclical, not seral, relationship with beech 'as part and parcel of the system'. In these forests small gaps are formed by death or windthrow. On most occasions birch, ash or oak (*Quercus* spp.) grow more vigorously than beech, fill the gap first and then are slowly overtopped and replaced by beech. Some even-aged stands of beech occur, having grown up together after a particularly favourable set of circumstances. These beechwoods are an extreme case: Britain has so few tree species that they form forests over a wider range of habitats than they would with more competition (Watt 1944). But the same processes can be seen at work in forests in many parts of the world.

Watt's elucidation of cyclic succession (or cyclic regeneration to use a more neutral term (Miles 1979)) included the first use of a forest profile diagram (Watt 1924, Fig. 3) and a study of seed germination and seedling establishment of beech and oak, which proved, as often, to be the most critical stages in determining future forest composition. A complex interplay was demonstrated between the different tree species and soil, microclimate, regional climate and biotic factors. For example, birch and beech on the Chilterns have different relationships depending on soil acidity (Watt 1919, 1923, 1924, 1926, 1934).

Virgin forests in the north temperate zone

The structure and reproduction of the virgin coniferous and deciduous forests of the north temperate zone were reviewed by Jones (1945), although he had virtually no data from China, a substantial lacuna which still remains to be plugged. Jones concluded that the climax forest probably fluctuates in space and time with different tolerant (shade-bearing) species replacing each other due to vagaries of reproduction and perhaps of site, but undisturbed by catastrophes for very long periods. Forests are also known where the same species perpetuates itself by recolonizing gaps in its own canopy. An example recently worked out in some detail is the high elevation balsam fir (*Abies balsamea*) forest of

---

* Nomenclature follows that of the papers cited.

north-east America (Sprugel 1976), where trees die at the leeward side of gaps which move progressively down-wind, passing through the forest about once every 60 yr.

The same view as Jones that the climax forest was that attained in the absence of catastrophes was also expressed by earlier workers in the forests of North America (see Raup 1964). It was thought that major disruptions began with European settlement. But it has become increasingly realized that hurricanes and fires (see Walker, this volume) have repeatedly destroyed many of these forests and extensive stands of intolerant species, not reproducing *in situ*, are a normal feature. General reviews have been published by Raup (1964), Bormann & Likens (1979), White (1979), Spurr & Barnes (1980) and Oliver (1981). Not all forests have been equally affected by such catastrophic disruptions and care is needed in extrapolating from one region to others. Fire was especially prevalent in the western Lake States and hurricanes in central New England. The White Mountain forests of New England appear to have had a low incidence of catastrophes and therefore had a high likelihood of reaching a stage of cyclic regeneration like that described for the English beechwoods. This stage has been termed a 'shifting mosaic steady state' by Bormann & Likens (1979). Since settlement man has greatly increased the extent and frequency of major disturbance so that today there are larger stands of even-aged young forests. In these North American forests continuing instability resulting from external, allogenic factors must be regarded as quite normal. In fact, even in Europe there are occasional extensive windstorms which destroy large tracts of forest (e.g. Wiebecke & Brunig 1975).

In conclusion, for the north temperate forest it appears realistic to take the broad view that climax forest includes patches of intolerant tree species as part of its mosaic.

Chilean forests

By way of a further example of the forest growth cycle and gap phase replacement we may consider recent detailed studies on certain temperate forests of south central Chile. These are a mosaic of communities, most of which are dominated by light-demanding species of *Nothofagus*, the southern beech. There are differences in species composition associated with latitude and altitude. It has been discovered that the various *Nothofagus* forests below about 1000 m elevation are seral, colonizers of the landslips, detritus avalanches, ash falls, lava- and mudflows associated with earthquakes and volcanic eruptions (Veblen & Ashton 1978). Major disturbances occur much more frequently than the five hundred years age attained by *Nothofagus* trees (Veblen *et al.* 1981). Small areas occur where *Nothofagus* species are dying out and being replaced by shade-bearing species. But most of the forests are dominated by *Nothofagus* and it seems reasonable to include them in any definition of climax forest in this region where cataclysms frequently occur.

Above 1000 m up to the timberline the more shade-tolerant species are absent leaving only *Nothofagus* spp. Here, too, tree regeneration may also depend on periodic disturbance. A bamboo (*Chusquea tenuiflora*) occurs which is commoner and more vigorous underneath evergreen *Nothofagus betuloides* and *N. dombeyi* than below deciduous *N. pumilio* and retards the establishment and growth of ground vegetation including tree seedlings. In the absence of massive exogenous disturbance tree regeneration must await synchronous flowering and death of the bamboo at intervals of 10–30 yr, its reduction by pests or disease, or its weakening beneath gaps following tree fall. These different processes result in the regeneration of patches of even-aged *Nothofagus* (Veblen 1979; Veblen, Veblen & Schlegel 1979; Veblen *et al.* 1981).

Ecological groups of tree species in temperate forests

In the limiting case of the British forests, very poor in tree species, a useful practical distinction can be made between intolerant, light-demanding pioneer species (e.g. birch), and tolerant or shade-bearing species (e.g. beech). Even here other aspects of their ecology blur this simple dichotomy. For example, beech is the first tree to grow on south-west slopes of the South Downs which are exposed to the prevailing winds (Watt 1924) and birch lives in cyclic succession with beech on the more acid soils of the plateau of the Chiltern Hills (Watt 1934). Moreover birch forms apparently stable forests with an open, light canopy at the extremities of its range in Britain and elsewhere in north-western Europe, as well as in certain subalpine situations. This seems to be because birch grows successfully beyond the tolerance range of species which elsewhere replace it in succession (Moore 1979). Birch and beech are at the extremes of the tolerance spectrum. In Britain, and even more so in continental Europe (Ellenberg 1978), other species lie between them.

The same temperate forest formation in North America has far more tree species than in Europe. Raup (1964) recorded thirty-eight deciduous and five coniferous native tree species on 1500 ha at Black Rock Forest in New York State. Foresters and ecologists have found it useful to recognize three (Bormann & Likens 1979) or even five different ecological groups, sometimes called tolerance classes (Baker 1950; Graham 1954; Spurr & Barnes 1980). The tolerance class of any individual species depends to some extent on site and varies in different parts of its range.

The regeneration requirements of codominant tree species in one rich Chilean forest were found to form a continuum of responses of different scales of disturbance, with the main canopy species *Aextoxicon punctatum* regenerating under continuous shade or in small openings at one extreme, and *Nothofagus obliqua* dependent on destruction of the forest stand at the other (Veblen, Ashton & Schlegel 1979).

PATTERN AND PROCESS IN SOME TROPICAL RAIN FORESTS

Tropical rain forests are structurally the most complex of all ecosystems. Amongst them are the most floristically rich ecosystems. There are commonly 50–150 tree species $\geqslant 0.3$ m girth per ha (Whitmore 1975, Fig. 1.5), and many more if smaller trees and other life-forms are added. Phenomena which are simplified in other forests are here seen at their greatest intricacy. Pattern is at its most diverse. It can be shown, however, that the same processes are at work as in other forests.

Some tropical rain forests are now known which are unstable, continually wracked by catastrophe, and with extensive stands of non-perpetuating tree species. Cyclones cause this devastation on northern Kolombangara in the Solomon Islands (Whitmore 1974) and on many of the Caribbean islands (e.g. Crow 1980). Major disturbance occurs to a variable degree. Landslides caused by earthquakes have been estimated to disturb 2% per century of the rain forests of the Darién region of Panama and 8–16% of Papua New Guinea (Garwood, Janos & Brokaw 1979). This latter territory also suffers volcanism and landslides caused by weathering.

By contrast, other rain forests are known in which the forest canopy is more stable, disturbed only by the small gaps caused by the death of an individual or a few trees and where replacement is by shade-bearing species. Most of the dipterocarp-dominated rain forest of Sumatra, Malaya, Borneo and the southern Philippines is like this. A similarly stable forest was described from Costa Rica by Hartshorn (1978) with gaps of average size only 0·09 ha, and from western Kolombangara by Whitmore (1974). Even in such forests extensive gaps form very occasionally. In Malaya and Borneo strong, extensive windstorms are so rare as to have excited particular attention (Whitmore 1975).

Ecological groups of tree species in tropical rain forests

Just as in temperate forests, useful generalizations can be made by dividing tropical rain forest tree species into ecological groups based on their response to light and hence to gaps in the forest canopy.

A class of pioneer species of open sites can clearly be identified. These have a whole set of characteristics, many of which are shared with their temperate forest equivalents, such as copious, small, readily dispersed seed and rapid seedling growth (Budowski 1963; Whitmore 1975). Seed is produced frequently if not continuously and there is accumulating evidence that in at least some tropical pioneer species it can remain dormant. Pioneer genera and many species are of wide geographical range. The main examples are *Cecropia* (70–80 spp.) and *Ochroma* (balsa, monotypic) in the neotropics, *Macaranga* (*c.* 80 spp) and *Musanga* (2 spp.) in Africa, *Macaranga* (*c.* 200 spp. but not all pioneers) in the eastern tropics and *Trema* (30 spp.) which is pantropical.

At the other extreme, each part of the humid tropics has a long list of genera and species which are shade-bearers, able to regenerate *in situ*. Examples are the legumes *Cynometra alexandri* in Africa and *Dicorynia guianensis* in South America and the 'heavy hardwood' class of dipterocarps in Malesia (e.g. *Neobalanocarpus* and *Shorea* section *Shorea*).

Furthermore, there is a broad class of species of intermediate tolerance, often known as late-secondary species, which cannot colonize bare open sites, but which become dominant at an intermediate stage of secondary succession and which do not perpetuate themselves *in situ*. This group contains most of the genera which provide timber of high commercial value. The neotropics have *Cedrela* (cigar box cedar) and *Swietenia* (true mahogany) as well as the lesser known *Bombacopsis* and *Cordia* and also *Dialyanthera*, *Iryanthera* and *Virola* of the Myristicaceae (Budowski 1965, 1970; Oldeman 1978). In Africa the main representatives are the African mahoganies *Entandrophragma*, *Guarea*, *Khaya* and *Loxoa* and also *Triplochiton*. In Asia a prominent example is the group of *Shorea* species which produce 'light hardwood' timber.

Closer examination has shown that species do not divide neatly into three groups in their requirement for light gaps for establishment. For example, the forests of Kolombangara Island have only twelve common big tree species. Yet these show a spectrum of responses and can be grouped into four classes (Whitmore 1974, 1975). The five *Piper* species (all small trees), present in a rain forest at Vera Cruz, Mexico, show a wide range of responses (Gómez Pompa 1971).

The forest growth cycle and forest succession

The biggest gaps in the forest are created by wide-scale destruction and the first trees to dominate belong to the most light-demanding species group, the pioneers. As time passes smaller gaps develop due, for example, to death or local wind-throw and species with a lesser light requirement grow up in them. Eventually the most shade-bearing species arrive. These have only a small light requirement, at least as juveniles, and can regenerate below themselves *in situ*. This progressive change is called secondary succession.

Examples are known of tropical rain forests where this sequential change is occurring. Floristic succession has been recorded in Nigeria where there are rain forests, about 200 yr old, dominated by the late seral, commercially valuable African mahoganies. These species are not regenerating but are being replaced by a self-perpetuating climax forest of fewer species which at present occur as small trees below the Meliaceae (Jones 1955–56). In Central America rain forests develop through four structural and three floristic phases, 'pioneer', 'late secondary' and 'climax' (whose characteristic genera were listed in the previous section) (Budowski 1965, 1970).

The concept of a steady succession in tropical rain forest from pioneer to climax (but liable to total or partial regression at any time) was elaborated by Oldemen (1978) and Hallé, Oldeman & Tomlinson (1978) as a complex model with five stages each floristically and structurally different. This paradigm seems too complex. No forest has yet been found which has so many sharply defined stages. The observed underlying generalities between forests seem in fact to be restricted to the forest growth cycle and the occurrence of species equipped to succeed in gaps of different size.

Apparently haphazard tree replacement rather than succession was postulated by Aubréville (1938; see also Eyre 1971) with the Ivory Coast rain forests in mind. This particular paper of a very prolific writer was used by Watt (1947) as an example of pattern and process; it became very well-known and acted as a stimulus to subsequent studies on rain forest dynamics (e.g. Jones 1955–56; Webb, Tracey & Williams 1972; Whitmore 1974), probably because Richards (1952) drew attention to it and termed the phenomena described the 'cyclical or mosaic theory of regeneration'. Aubréville described floristic composition as varying from place to place and at any one spot fluctuating in time. He did not make the distinction between early and late seral species and forests, though certain of his examples would now be so regarded.

STUDIES ON TREE AUTECOLOGY

These examples of forests in different parts of the world show how different tree species succeed in canopy gaps of different size. Further elucidation of forest dynamics comes from an understanding of the autecology of the different sorts of tree species. The ability of the seedling or sapling to survive under shade is the most important although not the only factor. To speak of light-demanding (or intolerant) and shade-bearing (or tolerant) species is a convenient shorthand.

Analysis of the ecophysiology of different tolerance classes is fragmentary (see reviews of Bazzaz 1979; Bazzaz & Pickett 1980). Light-demanding species often have higher rates of both photosynthesis and respiration. In deep shade rapid respiration, water stress, disease and predation collectively contribute to their death. Their success in full light may be partially attributed to architecture. North American examples investigated (Marks 1975; Bormann & Likens 1979) had indeterminate growth, continually added new leaves and recirculated nutrients to them. They had smaller, shallower root systems and showed a quick response to fertilizer but were less able to compete for water and nutrients.

Gap size

There are major differences in microclimate between big and small gaps; within any one gap conditions differ greatly from the edge to the centre, the more so in big gaps. Gap shape also has an influence (Oldeman 1978). Several studies in tropical forests have investigated the size of gap at which light-demanding rather than shade-bearing species successfully establish.

The classic investigation was on Mount Gede in west Java. Artificial clearings of 1000 m^2 were soon filled by surviving primary forest trees but in clearings of 2000 m^2 and 3000 m^2 these were swamped by invading pioneers (Kramer 1933). More recent studies have added little to this discovery. Most have been of small gaps in which dryads were establishing. In Surinam it was found that dryads succeeded in clearings of 100–1000 m^2, but in larger clearings secondary succession took place (Schulz 1960). In a Costa Rican forest small natural gaps of 900 m^2 were all filled by dryads (Hartshorn 1978). In one Ivory Coast forest dryads succeeded in small clearings of 100 m^2 (Nierstrasz 1975). In another Ivory Coast forest a more complex situation prevailed. Here there was no floristic difference in the regrowth between gaps in the range 50–500 m^2. Most gaps had some individuals of the pioneer tree *Macaranga barteri* and there were more in the larger ones. But all gaps differed from forest regrowing on large cleared areas in lacking the late-secondary group of species (Bonnis 1980). Hartshorn (1980) reported that light and high surface temperatures necessary to break the dormancy of balsa occur in gaps of over 400 m^2.

The general conclusion from these scattered observations is that conditions for the establishment of a forest of pioneers commonly occur in gaps of over about 1000 m^2.

Source of replacement trees during regeneration

New tree stems in a gap grow up from roots, stumps or fallen trunks, from seedlings or from seed.

The seed may either be lying dormant in the soil as a 'seed-bank' or be conveyed into the gap as 'seed rain'. For most tropical forests we do not yet know which of these conditions prevails. The early evidence that rain forest soil contains seeds of pioneers (Symington 1933; Keay 1960) did not demonstrate that the seed had lain dormant since it could have recently arrived. Recently, however, a tree seed bank, accumulated over a period, has been unequivocally demonstrated in the soil under a tropical forest in Thailand (Cheke *et al.* 1979). Germination may be triggered by a variety of stimuli (Bazzaz & Pickett 1980). High soil temperature is necessary for balsa (Vázquez-Yanes 1974) and altered light quality for, e.g. *Piper hispidum* (Wiechers & Vázquez-Yanes 1974).

Some temperate light-demanders also accumulate a soil seed store, e.g. the pin cherry (*Prunus pennsylvanica*) of the north-eastern American deciduous forest (Marks 1974). By contrast the birches, many of which are pioneers, do not appear to have a soil seed store. Their successful occupation of a gap depends on seed rain, and on suitable surface conditions for germination and establishment (Sarvas 1948; Marquis 1969).

Sequential or simultaneous arrival

A large gap may be occupied by pioneer species and later colonized by shade-bearers which eventually supersede, so that there is a floristic succession, sometimes referred to as 'relay floristics'. Alternatively species of the various ecological classes may arrive more or less simultaneously but successively attain dominance, the faster growing, shorter lived pioneers being eventually overtopped and replaced (Egler 1954; Drury & Nisbet 1973;

Oliver 1981). Examples are known of both possibilities. Variants combining both extremes are explored by Noble & Slatyer (1979). The matter has obvious implications for silviculture.

In the tropical dipterocarp rain forest at Sungai Kroh, Malaya, and on Vanikolo Island seedlings of dryads which survived forest clearance (*Shorea* spp. and *Agathis macrophylla*) died out of the pioneer forest stand and only later re-invaded (Kochummen 1966; Whitmore 1966). But in the regrowth of rain forest on abandoned fields on Mindanao both pioneer trees and later-successional tree ferns (*Cyathea*) were found present together with the initial herbaceous weed vegetation (Kellman 1972).

The regrowth in small gaps in tropical rain forest is characteristically by seedlings which established under the previous canopy and persisted awaiting such 'release'. In big gaps these seedlings die sooner or later. This is well shown by many Dipterocarpaceae which also exhibit differences between species in the frequency of seed production, hence recruitment to the seedling population, and in seedling longevity (Whitmore 1975).

In the northern hardwood forest of North America studied by Bormann & Likens (1979) most of the species and individuals of early regrowth were present in the intact forest at the time of clear-cutting and the others entered soon after. Also present were most of the individuals which would ultimately become the dominant trees. As this forest develops different species dominate in sequence dependent on their different growth characteristics. Only after about a century do new individuals enter. Forest succession on land which has been cultivated is however different. Here most tree species enter the site gradually, in a sequence fairly predictable from their frequency of good seed years and powers of dispersal.

SPECIES RICHNESS AND FOREST STABILITY

Forests, then, have a spatial pattern, or mosaic, of patches of different structure and in most forests also of floristic composition. We can therefore gain an insight into both species richness and stability by analysing forests in terms of pattern and process. This insight is valuable for the design of conservation areas.

Species richness

The extreme floristic richness of many tropical rain forests has given rise to speculation on its evolutionary origin and ecological maintenance (Fedorov 1966; Ashton 1969, 1977). In fact both species richness and forest diversity vary at a range of scales which need to be disentangled (Whitmore 1975). At the highest level is the availability of flora. This depends upon the way evolution has progressed and on past extinctions. For example, by the chances of evolution, most formations of the tropical rain forests of west Malesia are dominated by the single family, Dipterocarpaceae. So far as is known these are the only forests in which one tree family achieves such a high density of genera, species and individuals.

The European temperate deciduous forest is impoverished in comparison with the East Asian and North American because of Pleistocene extinctions during glacial maxima.

Tropical forests were for long believed to have been stable and to owe their species richness in part to historical stability (e.g. Ashton 1969; Stebbins 1974) but it is now believed that they too were affected by Pleistocene climatic change and that at the present day rain forest is at or near its maximum extent, which it has only attained for a small fraction of the last two million years (Flenley 1979; Prance 1981). The ranges of species and the greater richness of some areas than others are in part a result of Pleistocene climatic fluctuations.

At a smaller scale forests differ in species richness with site. For example the heath forest, peatswamp forest and lowland evergreen forest formations of Borneo lie in close proximity and have substantial differences.

At a still lower scale variation occurs within each forest formation. Part of it relates to topography. For example, there are structural and floristic differences in the lowland evergreen rain forest at Andulau, Brunei, between ridges, hillsides and valleys (Ashton 1964). It is at a yet lower level that forest dynamics play a major role in determining species composition because some species are adapted to regenerate in big gaps and others in small. The latter species depend on conditions created by the former. They fit into the frame of a forest already formed, as part of its growth cycles. Species richness is maintained by continuing disturbance at several scales.

It has been postulated that the forests richest in species are those recovering from a major disturbance because they contain both the pioneer trees and their successors. This 'intermediate disturbance hypothesis' (Connell 1978, 1979), which was expressed earlier in more general terms by Margalef (1963), is supported by observations on the forests of the eastern United States (Spurr & Barnes 1980, p. 445) and on some tropical rain forests, namely the Nigerian rain forest (described above) and the Ugandan rain forest outliers in which the climax is a consociation dominated by self-perpetuating *Cynometra alexandri* which is replacing a more species-rich forest containing many Meliaceae (Eggeling 1947). In addition, the Luquillo rain forest of Puerto Rico, recovering from a 1943 hurricane, has become floristically poorer with time (Crow 1980), and Budowski (1970) cites other examples from Costa Rica and Panama.

But there is no *a priori* reason why late secondary forests should be richest in species. Within a single forest formation several examples are known where Connell's hypothesis does not hold. Richness in fact depends on the number of species in the region under discussion (i.e. the available flora) which are adapted to each facet of the regeneration niche. For example, in the western Solomon Islands there are in total more tree species requiring shade or small gaps to establish than there are big-gap colonizers. Observations suggest that the forests most rich in tree species (occurring in patches of *c.* 200 km² each) are those not under the influence of major disturbance and with a fine mosaic of structural phases (Whitmore 1974 and unpublished). Moreover, the picture is complicated if plants of the dependent synusiae are added to the trees. In the western Solomons, as secondary succession progresses, the number of climber species decreases and of epiphytes increases (Whitmore 1974). The forests of Barro Colorado Island, Panama, show a rapid increase in species richness to 15 yr age then a slower increase to 65 yr. Forests 130 yr old are still changing (Knight 1975). Here too, species richness is not maximal at the late secondary stage.

Niche width in tropical rain forest

There are very large numbers of tree species on small plots of most tropical rain forests (50–150/ha, see above) and the number increases as plot size is increased or minimum girth for inclusion is decreased (see e.g. Whitmore 1975, Figs. 1.3–1.5). In the Pasoh (Malaya) and Andulau (Brunei) forests local variation in species composition arises from the association of groups of species partially with slight variations in relief and soil and partially with the forest growth cycle. At this level the patterns are very subtle and the results obtained have been coloured by the sampling and analytical procedure used (Wong & Whitmore 1970; Austin, Ashton & Greig-Smith 1972; Whitmore 1975; Ashton 1976).

The outstanding question is whether this great richness in species is dependent on each having a different niche, as favoured by Ashton (1977) and as Gause's competitive exclusion principle demands, or whether many species have almost identical niches. There may be several species more or less equally equipped to succeed in a gap of any given size. Connell (1979) has pointed out that it is in fact impossible to test the niche-specialization hypothesis for although 'with enough imagination environmental heterogeneity can be used to account for all diversity . . . how can one decide with certainty if species are adequately specialized, or one has looked at correct niches?' It remains the case that there have been very few autecological studies of rain forest trees, attempting to define the dimensions of their niche: Whitmore (1966) on *Agathis macrophylla*, Boaler (1966) on *Pterocarpus angolensis*, Lee (1967) and Burgess (1969) on the dipterocarps *Dryobalanops aromatica* and *Shorea curtisii* respectively.

Even if allowance is made for all the known differences of possible ecological importance it is difficult to believe that they are enough to fit as many niches as required by Gause's principle. One is forced to conclude that although there are indeed numerous differences between species in numerous facets of their ecology, many tree species in species-rich forests have largely overlapping or almost identical niches. In terms of regeneration of these forests, this means that any one of several species is well-equipped to grow in a gap of any particular size and above- and below-ground microclimate, and there is an important element of chance in determining which one succeeds (Richards 1969; Whitmore 1975). Chance operates at many levels, for example, on which species of several happens to be either present as a seedling below a gap when it forms or reaches it and grows up first.

Stability

We have seen that no forest is stable in the sense of being unchanging. All are in a continuous state of flux. The forest growth cycle causes patchiness in structure and patch size has an important influence on species composition. Within the most species-rich forests it seems likely that even patches of the same size each only contain a selection of the suitably equipped species. As a forest recovers from major disturbance a coarse mosaic gives way to a finer one, until the whole is set back by the next catastrophe, so greater stability in the sense of less change increases with succession. General accounts have recently been published by Horn (1974, 1976), Connell & Slatyer (1977) and Whittaker & Levin (1977).

Tropical rain forests were considered to be 'fragile' by Gómez Pompa, Vázquez-Yanes & Guevara (1972). But this has sometimes been misinterpreted in a wider sense than intended, e.g. by Farnworth & Golley (1974). It is true in some respects. Forest *qua* forest is robust, especially in the humid tropics where all clearings, except those with very serious soil erosion or soil compaction by heavy machinery, very quickly grow a new forest. The dynamic flux of the forest canopy is not unique or different in the tropics. Extensive disturbance, especially where it is soon repeated, does however lead to floristic impoverishment. In this sense these forests are indeed fragile. Many of the late seral and climax tree species of apparently equivalent or nearly equivalent ecological niche occur at low population densities. Many produce fruit infrequently, have seeds with no dormancy and poor dispersal and depend for their perpetuation on an adequate occurrence of the right regeneration niche. Dipterocarpaceae fit this description.

CONCLUSIONS

Scientific ecology is a young subject almost entirely a product of the present century. Empirical forest ecology was developed earlier by silviculturists because of its practical importance. The early view of forests taken by ecologists was coloured by the concept of succession to climax. Evidence of continuing disturbance was played down. Now the emphasis has changed. Disturbance at many scales is recognized to be always present and it is generally felt unrealistic to disregard it. Instead of progression to a stable endpoint, disturbances, whether arising within the community (autogenic) or outside it (allogenic) (Webb, Tracey & Williams 1972), are regarded as normal. Hence unambiguous generalizations about succession are difficult to make (Drury & Nisbet 1973; White 1979; Oliver 1981). Such a view of the forest is more consistent with the repeatedly demonstrated physical instability of habitat, much of it unpredictable.

Ecosystems at different stages of development have different values to mankind (Odum 1969). Forests in the building phase of the growth cycle have the greatest productivity. In forestry the product sought determines the age at which the forest is economically mature. For fuelwood, or chemical feedstock, this may be early building phase, for chips and poles it is the building phase at the moment when productivity (current annual wood production) begins to drop and for high quality timber it is somewhere early in the mature phase. Where conservation values prevail mature phase forest and a range of patches of all sizes are all needed, for this maximizes ecosystem and hence niche diversity and species richness of plants (Pickett & Thompson 1978; Connell 1978, 1979; Gilbert 1980) as well as animals (Wiens 1976).

Thus, in the manipulation of forest ecosystems, just as in understanding their nature, it can be seen that by 'studying relationships in space and time (how it is made and how it works) we may, with sufficient understanding, devise a working model which can be used as a guide to action', and that 'the dynamic approach has a rich reward' (Watt 1944, 1964).

ACKNOWLEDGMENTS

My attempt to grapple with this subject was originally stimulated by a suggestion made to me by Dr A. S. Watt in about 1969 that it would be interesting to explore the influence of foresters as empirical ecologists. I am grateful to the following for comments: D. McC. Newbery, E. I. Newman, R. A. A. Oldeman, J. Proctor, E. R. C. Reynolds and T. T. Veblen.

REFERENCES

Ashton, P. S. (1964). Ecological studies in the mixed dipterocarp forests of Brunei State. *Oxford Forest Memoirs*, **25**.
Ashton, P. S. (1969). Speciation among tropical forest trees: some deductions in the light of recent evidence. *Biological Journal of the Linnean Society*, **1**, 155–196.

Ashton, P. S. (1976). Mixed dipterocarp forest and its variation with habitat in the Malayan lowlands: a re-evaluation at Pasoh. *Malaysian Forester*, **39**, 56–72.

Ashton, P. S. (1977). A contribution of rain forest research to evolutionary theory. *Annals of the Missouri Botanic Garden*, **64**, 649–705.

Aubréville, A. (1938). La Forêt coloniale: les forêts de l'Afrique occidentale française. *Annales Académie Sciences Coloniale*, **9**, 1–245.

Austin, M. P., Ashton, P. S. & Greig-Smith, P. (1972). The application of quantitative methods to vegetation survey. III. A re-examination of rain forest data from Brunei. *Journal of Ecology*, **60**, 305–324.

Baker, F. S. (1950). *Principles of Sylviculture*. McGraw Hill, New York.

Bazzaz, F. A. (1979). The physiological ecology of plant succession. *Annual Review of Ecology and Systematics*, **10**, 351–371.

Bazzaz, F. A. & Pickett, S. T. A. (1980). Physiological ecology of tropical succession: a comparative review. *Annual Review of Ecology and Systematics*, **11**, 287–310.

Boaler, S. B. (1966). The ecology of *Pterocarpus angolensis* DC. in Tanzania. *Overseas Research Publication, Ministry of Overseas Development, London*, 12.

Bonnis, G. (1980). *Etude des chablis en forêt dense humide de sempervirente naturelle de Taï (Côte d'Ivoire)*. ORSTOM, Abidjan.

Bormann, F. H. & Likens, G. E. (1979). *Pattern and Process in a Forested Ecosystem*. Springer, New York.

Budowski, G. (1963). Forest succession in tropical lowlands. *Turrialba*, **13**, 42–44.

Budowski, G. (1965). Distribution of tropical American rain forest species in the light of successional processes. *Turrialba*, **15**, 40–42.

Budowski, G. (1970). The distinction between old secondary and climax species in tropical central American lowland forests. *Tropical Ecology*, **2**, 44–48.

Burgess, P. F. (1969). Preliminary observations on the autecology of *Shorea curtisii* Dyer ex King in the Malay Peninsula. *Malayan Forester*, **32**, 438.

Cheke, A. S., Weerachal Nanakorn & Chusee Yankoses (1979). Dormancy and dispersal of seeds of secondary forest species under the canopy of a primary tropical rain forest in northern Thailand. *Biotropica*, **11**, 88–95.

Connell, J. H. (1978). Diversity in tropical rain forests and coral reefs. *Science*, **199**, 1302–1310.

Connell, J. H. (1979). Tropical rain forests and coral reefs as open non-equilibrium systems. *Population Dynamics* (Ed. by R. M. Anderson, B. D. Turner & L. R. Taylor), pp. 141–163. Blackwell Scientific Publications, Oxford.

Connell, J. H. & Slatyer, R. O. (1977). Mechanisms in natural communities and their role in community stability and organisation. *American Naturalist*, **111**, 1119–1145.

Cousens, J. (1974). *An Introduction to Woodland Ecology*. Oliver and Boyd, Edinburgh.

Crow, T. R. (1980). A rain forest chronicle: a 30 year record of change in structure and composition at El Verde, Puerto Rico. *Biotropica*, **12**, 42–55.

Drury, W. H. & Nisbet, I. C. T. (1973). Succession. *Journal of the Arnold Arboretum*, **54**, 331–368.

Eggeling, W. J. (1947). Observations on the ecology of the Budongo rain forest, Uganda. *Journal of Ecology*, **34**, 20–87.

Egler, F. E. (1954). Vegetation science concepts. I. Initial floristic composition, a factor in old field vegetation development. *Vegetatio*, **4**, 412–417.

Ellenberg, H. (1978). *Vegetation Mitteleuropas mit den Alpen*. Ulmer, Stuttgart.

Eyre, S. R. (1971). Regeneration patterns in the closed forest of Ivory Coast. Chapter 2 (Translation of paper by A. Aubréville).

Farnworth, E. G. & Golley, F. B. (Eds) (1974). *Fragile Ecosystems*. Springer, New York.

Fedorov, An. A. (1966). The structure of the tropical rain forest and speciation in the humid tropics. *Journal of Ecology*, **54**, 1–11.

Flenley, J. R. (1979). *The Equatorial Rain Forest: a Geological History*. Butterworth, London.

Garwood, N. C., Janos, D. P. & Brokaw, N. (1979). Earthquake-caused landslides: a major disturbance to tropical forests. *Science*, **205**, 997–999.

Gilbert, L. E. (1980). Food web organisation and conservation of neotropical diversity. *Conservation Biology* (Ed. by M. E. Soule & B. A. Wilcox). Sinauer, Sunderland, Mass.

Gómez-Pompa, A. (1971). Posible papel de la vegetación secundaria en la evolución de la flora tropical. *Biotropica*, **3**, 125–135.

Gómez-Pompa, A., Vázquez-Yanes, C. & Guevara, S. (1972). The tropical rain forest: a non-renewable resource. *Science*, **177**, 762–765.

Graham, S. A. (1954). Scoring tolerance of forest trees. *Michigan Forestry Vol. 4*, University of Michigan, Ann Arbor.

Haeckel, E. H. (1869). Entwicklungsgang und Aufgabe der Zoologie. *Jenaische Zeitschrifte für Naturwissenschaft*, **5**, 353–370.

Hallé, F., Oldeman, R. A. A. & Tomlinson, P. B. (1978). *Tropical Trees and Forests*. Springer, New York.

Hartshorn, G. S. (1978). Tree falls and tropical forest dynamics. *Tropical Trees as Living Systems* (Ed. by P. B. Tomlinson & M. H. Zimmermann), pp. 617–638. University Press, Cambridge.

Hartshorn, G. S. (1980). Neotropical forest dynamics. *Biotropica*, **12** (Suppl.), 23–30.

Horn, H. S. (1974). The ecology of secondary succession. *Annual Review of Ecology and Systematics*, **5,** 25–37.

Horn, H. S. (1976). Succession. *Theoretical Ecology* (Ed. by R. M. May), pp. 187–204. Blackwell Scientific Publications, Oxford.

Jones, E. W. (1945). The structure and reproduction of the virgin forest of the north temperate zone. *New Phytologist*, **44,** 130–148.

Jones, E. W. (1955–56). Ecological studies on the rain forest of southern Nigeria. IV. The plateau forest of the Okumu forest reserve. *Journal of Ecology*, **43,** 564–594; **44,** 83–117.

Keay, R. W. J. (1960). Seeds in forest soil. *Nigerian Forestry Information Bulletin* (new series), **4,** 1–12.

Kellman, M. C. (1972). The successional role of tree ferns in a tropical montane environment. *International Geography* (Ed. by W. P. Adams & F. M. Helleiner), Vol. 1, pp. 264–266. University of Toronto Press, Toronto.

Knight, D. H. (1975). A phytosociological analysis of species-rich tropical forest on Barro Colorado Island, Panama. *Ecological Monographs*, **45,** 259–284.

Kochummen, K. M. (1966). Natural plant succession after farming at Sungei Kroh. *Malayan Forester*, **29,** 170–181.

Kramer, F. (1933). De natuurlijke verjonging in het Goenoeng Gedeh complex. *Tectona*, **26,** 156–185.

Lee, P. C. (1967). *Ecological studies on* Dryobalanops aromatica *Gaertn. f.* Thesis, University of Malaya.

Margalef, R. (1963). On certain unifying principles in ecology. *American Naturalist*, **97,** 357–374.

Marks, P. L. (1974). The role of pin cherry (*Prunus pennsylvanica* L.) in the maintenance of stability in northern hardwood ecosystems. *Ecological Monographs*, **44,** 73–88.

Marks, P. L. (1975). On the relation between extension growth and successional status of deciduous trees of the northeastern United States. *Bulletin of the Torrey Botanical Club*, **102,** 172–177.

Marquis, D. A. (1969). Silvical requirements for natural birch regeneration. *Birch Symposium Proceedings* (Ed. by E. Larsen), pp. 40–49. United States Department of Agriculture, Forest Service N.E. Forest Experiment Station.

Miles, J. (1979). *Vegetation Dynamics.* Chapman and Hall, London.

Moore, P. D. (1979). Next in succession. *Nature*, **282,** 362–363.

Nierstrasz, E. (1975). *Clairières et chablis en forêt naturelle.* ORSTOM, Centre d'Adiopodoume, Mimeo.

Noble, I. R. & Slatyer, R. P. (1979). The use of vital attributes to predict successional change in plant communities subject to recurrent disturbance. *Vegetatio*, **43,** 5–21.

Odum, P. (1969). The strategy of ecosystem development. *Science*, **164,** 262–279.

Oldeman, R. A. A. (1978). Architecture and energy exchange of dicotyledonous trees in the forest. *Tropical Trees as Living Systems* (Ed. by R. B. Tomlinson & M. H. Zimmermann), pp. 535–560. University Press, Cambridge.

Oliver, C. D. (1981). Forest development in North America. *Forest Ecology and Management*, **3,** 153–168.

Pickett, S. T. A. & Thompson, J. N. (1978). Patch dynamics and the design of nature reserves. *Biological Conservation*, **13,** 27–37.

Prance, G. T. (Ed.) (1981). *Biological Diversification in the Tropics.* Columbia University Press, New York.

Raup, H. M. (1964). Some problems in ecological theory and their relation to conservation. *Journal of Ecology*, **52** (Suppl.), 19–28.

Richards, P. W. (1952). *The Tropical Rain Forest.* University Press, Cambridge.

Richards, P. W. (1969). Speciation in the tropical rain forest and the concept of the niche. *Biological Journal of the Linnean Society*, **1,** 149–154.

Sarvas, R. (1948). Summary: A research on the regeneration of birch in South Finland. *Communicationes Instituti Forestalis Fenniae*, **35,** 82–91.

Schulz, J. P. (1960). *Ecological Studies on the Rain Forest in Northern Suriname.* North Holland, Amsterdam.

Sprugel, D. G. (1976). Dynamic structure of wave-generated *Abies balsamea* forests in the north-eastern United States. *Journal of Ecology*, **64,** 889–911.

Spurr, S. H. (1952). Origin of the concept of forest succession. *Ecology*, **33,** 426–427.

Spurr, S. H. & Barnes, B. V. (1980). *Forest Ecology*, 3rd edn. John Wiley, New York.

Stebbings, G. L. (1974). *Flowering Plants: Evolution above the Species Level.* Edward Arnold, London.

Steenis, C. G. G. J. van (1958). Rejuvenation as a factor for judging the status of vegetation types. The biological nomad theory. *Proceedings of the Symposium on Humid Tropics Vegetation, Kandy.* UNESCO, Paris.

Symington, C. F. (1933). The study of secondary growth on rain forest sites in Malaya. *Malayan Forester*, **2,** 107–117.

UNESCO/UNEP/FAO (1978). *Tropical Forest Ecosystems.* UNESCO, Paris.

Vázquez-Yanes, C. (1974). Studies on the germination of seeds: *Ochroma lagopus* Swartz. *Turrialba*, **24,** 176–179.

Veblen, T. T. (1979). Structure and dynamics of *Nothofagus* forests near timberline in south-central Chile. *Ecology*, **60,** 937–945.

Veblen, T. T. & Ashton, D. H. (1978). Catastrophic influences on the vegetation of the Valdivian Andes, Chile. *Vegetatio*, **36**, 149–167.

Veblen, T. T., Ashton, D. H. & Schlegel, F. M. (1979). Tree regeneration strategies in a lowland *Nothofagus*-dominated forest in south-central Chile. *Journal of Biogeography*, **6**, 329–340.

Veblen, T. T., Veblen, A. T. & Schlegel, F. M. (1979). Understorey patterns in mixed evergreen-deciduous forests in Chile. *Journal of Ecology*, **67**, 809–823.

Veblen, T. T., Donoso, Z. C., Schlegel, F. M. & Escobar, R. B. (1981). Forest dynamics in south-central Chile. *Journal of Biogeography*, **8**, 211–247.

Watt, A. S. (1919). On the causes of failure of natural regeneration in British oakwoods. *Journal of Ecology*, **7**, 173–203.

Watt, A. S. (1923). On the ecology of the British beechwoods with special reference to their regeneration. Part I. The causes of failure of natural regeneration of the beech (*Fagus silvatica* L.). *Journal of Ecology*, **11**, 1–48.

Watt, A. S. (1924–5). On the ecology of British beechwoods with special reference to their regeneration. Part II. The development and structure of beech communities on the Sussex Downs. *Journal of Ecology*, **12**, 145–204; **13**, 27–73.

Watt, A. S. (1926). Yew communities of the South Downs. *Journal of Ecology*, **14**, 282–316.

Watt, A. S. (1934). The vegetation of the Chiltern Hills with special reference to the beechwoods and their seral relationship. Part I. *Journal of Ecology*, **22**, 230–270; Part II. *Ibid*, **22**, 445–507.

Watt, A. S. (1944). Ecological principles involved in the practice of forestry. *Journal of Ecology*, **32**, 96–104.

Watt, A. S. (1947). Pattern and process in the plant community. *Journal of Ecology*, **35**, 1–22.

Watt, A. S. (1964). The community and the individual. *Journal of Ecology*, **52** (Suppl.), 203–212.

Webb, L. J., Tracey, J. G. & Williams, W. T. (1972). Regeneration and pattern in the subtropical rain forest. *Journal of Ecology*, **60**, 675–695.

White, P. S. (1979). Pattern, process and natural disturbance in vegetation. *Botanical Review*, **45**, 229–299.

Whitmore, T. C. (1966). The social status of *Agathis* in a rain forest in Melanesia. *Journal of Ecology*, **54**, 285–301.

Whitmore, T. C. (1974). Change with time and the role of cyclones in tropical rain forest on Kolombangara, Solomon Islands. *Commonwealth Forestry Institute Paper 46*.

Whitmore, T. C. (1975). *Tropical Rain Forests of the Far East*. Clarendon Press, Oxford.

Whitmore, T. C. (1978). Gaps in the forest canopy. *Tropical Trees as Living Systems*. (Ed. by P. B. Tomlinson & M. H. Zimmermann), pp. 639–655. University Press, Cambridge.

Whittaker, R. H. & Levin, S. A. (1977). The role of mosaic phenomena in natural communities. *Theoretical Population Biology*, **12**, 117–139.

Wiebecke, C. & Brunig, E. F. (1975). Education for world forestry. *Education*, **12**, 53–63.

Wiechers, B. L. & Vázquez-Yanes, C. (1974). Germinacion de semillas de *Piper hispidum* bajo diferentes condiciones de iluminación. *Investigaciones Sobre la Regeneración de Selvas Altas en Veracruz, Mexico*. (Ed. by A. Gómez Pompa, S. del Amo Rodríguez, C. Vázquez-Yanes & A. Butanda Cervera). Compañia Editorial Continental, Mexico.

Wiens, J. A. (1976). Population responses to patchy environments. *Annual Review of Ecology and Systematics*, **7**, 81–120.

Wong, Y. K. & Whitmore, T. C. (1970). On the influence of soil properties on species distribution in a Malayan lowland dipterocarp rain forest. *Malayan Forester*, **33**, 42–54.

NICHE SEPARATION AND SPECIES DIVERSITY IN TERRESTRIAL VEGETATION

E. I. NEWMAN

Department of Botany, University of Bristol, Bristol BS8 1UG, U.K.

SUMMARY

This paper considers the part which niche separation, related to variation in the habitat, plays in controlling diversity of terrestrial plant communities. It considers in particular the question: 'How different do species need to be in order to coexist?'.

The paper describes two mathematical models of plant communities in which niche separation occurs in the regeneration phase, i.e. germination or seedling establishment, but the adults compete without niche separation. One model applies when there are separate habitat states (e.g. disturbed *v.* undisturbed ground), the other when there is continuous habitat variation (e.g. size of microsites for seeds).

The models illustrate how the number of species coexisting and their relative abundance can depend not only on the abundance of the different habitat states which favour regeneration of each species but also on the balance of competition in the adult stage. Minimum differences between species are predicted, below which they are unlikely to coexist. When applied to real examples these suggest that usually only two or three species can achieve adequate niche separation along one habitat axis.

INTRODUCTION

Diversity in ecosystems has long fascinated ecologists, who have asked such questions as: 'How can so many species coexist, without some excluding others?' and 'Why are there more species in some ecosystems than others?'. A. S. Watt has made a major contribution to our understanding of diversity in vegetation, particularly by showing how plants themselves can impose structure and diversity in a plant community (Watt 1947). It therefore seems appropriate in this symposium to try to advance a little further our understanding of the control of diversity in vegetation, though my approach will be very different from Watt's.

It is generally accepted that in order to coexist more than transiently species must differ—they must show niche separation. If species are too similar all but one will be eliminated in competition. This paper considers the question 'How different do plant species need to be to coexist?'. The answer to this question would be a major step towards understanding the control of diversity. Animal ecologists have already provided some answers to the question (MacArthur & Levins 1967; MacArthur 1972; May 1974). To illustrate these authors' approach, suppose that in Fig. 4 A–F are six bird species each of which eats seeds of sizes indicated by the frequency distribution curves (ignoring dashed lines). MacArthur & Levins (1967) produced a model of three species competing for a single resource (e.g. seeds) which predicted that the three species would coexist indefinitely if $d/w > 1.6$ (MacArthur 1972). For the meaning of d and w see Fig. 4. May (1974) produced a model which could have any number of species and took into account

0262–7027/82/0300–0061$02.00 © 1982 British Ecological Society

fluctuations in the environment. It predicted coexistence of many species if $d/w > 1$, but a modified version gives 1·4 as the limiting value (Ågren & Fagerström 1980). These models provide simple and usable answers to the question of how different species need to be, or how heterogeneous the environment needs to be, to allow coexistence, and they have been applied to examples among animals (May 1974). It is tempting to apply the same rules to plants, and some workers have done so, e.g. May (1974, Fig. 6.8) and Platt & Weis (1977). However, there are several reasons why alternative models would be more appropriate for plants.

(1) The MacArthur and May models are based on the Lotka–Volterra competition equations. These assume that the number of individuals of a species is, on its own, adequate to describe the population; they were designed to examine population variation through time, and the importance of the potential rate of reproduction. All these features are appropriate for studying animals. However, in most perennial vegetation, while the size of individual plants may vary greatly the total biomass per unit ground area is relatively stable. The seed production per plant is usually so large that any gap caused by death of a mature plant is soon occupied by seedlings.

(2) The MacArthur and May models assume that the species compete for only a single resource. Even extensions of the models to two or three dimensions (Yoshiyama & Roughgarden 1977; Rappoldt & Hogeweg 1980) assume that niche separation occurs in each dimension. Among plants, however, there is likely to be competition for at least one requirement which does not provide niche separation. In particular, competition for light is very common, yet it probably rarely provides niche separation. Critical evidence that plants are competing for light has not often been obtained, but the fact that leaves of different plants so often overlap, and light intensity reaching lower leaves is so often reduced, is sufficient to show that competition for light is widespread. If, for example, two species differ in their requirements for seed germination and this provides niche separation, to ensure their coexistence this must be sufficient to counteract the competitive advantage of one species over the other in the vegetative phase, where plants would almost certainly compete for one resource and quite probably for more than one.

(3) Although many examples of niche separation in plants can be considered as a continuous resource axis, e.g. root depth distribution or response of seed production to a weather factor, many others cannot: for example, various sorts of disturbance either happen or do not happen, and one can still ask how different species need to be in response to these qualitative differences in the environment.

The models described in this paper are designed to take account of these features of plant communities.

ESSENTIAL FEATURES OF THE MODELS

The models concern the coexistence of species in stable vegetation. By 'coexist' I mean that individuals of the species compete with each other but no species is eliminated over a period of centuries. By 'stable' I mean that although the environment and vegetation may fluctuate from year to year there is no major large-scale disturbance, such as the fires considered by Walker (this volume). Small scale disturbance, e.g. a cattle hoof-print in grassland or a gap caused by the death of a tree in a forest, I regard as part of the dynamics within the stable community (see also Whitmore, this volume). In the models the species differ in their response to factors of the physical environment, e.g. soil or weather. Specifically excluded are situations where particular species themselves impose heterogeneity on the environment. This can happen because one species provides physical protection for another against wind or grazing, because each species alters its environment chemically, e.g. by toxic exudates, or because adult plants act as sources of inoculum for specific fungi or insects which attack their seeds or seedlings. Such situations require different models. The niche separation considered here can be in either space or time.

(1) *Separation in space.* Plants whose shoots intermingle may have roots developed at different depths (e.g. Weaver 1919). Roots of different species may occupy different pore sizes (Sheikh & Rutter 1969). Harper, Williams & Sagar (1965) provided some evidence that seeds, by virtue of their different size or shape, may germinate best in different 'safe sites' in soil into which individual seeds fit. There may be horizontal variation within a single plant community in physical or chemical properties of the soil or in factors of the aerial environment (e.g. light intensity on a forest floor). Local variation caused by animals (e.g. ant-hills, cattle hoof-prints) or physical disturbance (e.g. wind throw) can be included, provided the frequency and characteristics of gaps are controlled mainly by the disturbing agent rather than by the particular species of plant previously present.

(2) *Separation in time.* Separation within the year occurs, for example, between winter- and summer-germinating desert ephemerals, herbaceous forest species which emerge earlier than the trees above them, codominants which take up nutrients at different seasons (Rogers & Westman 1979). Variation from year to year in seed production, germination or seedling establishment could also provide opportunities for niche separation.

I shall refer to 'niches' and 'habitat states'. A niche is the range of habitat conditions in which a species occurs; a habitat state is a set of habitat conditions, in which one or more species may occur, and which is usually less extensive than the niche of a species. Examples of a habitat state could be soil of pH 5·6–6·0, years with August rainfall between 50 and 75 mm, cattle hoof-prints.

In these models the species compete in the vegetative state without niche separation, but there is niche separation in seed production, germination or seedling establishment. This seems appropriate for study: Grubb (1977) cited many examples of how species differ in the response of their seed production, germination and seedling establishment to environmental factors, but he failed to show whether these differences would permit the species to coexist. In the models the total number of plants is fixed: when an adult plant dies it is rapidly replaced by another, though not necessarily of the same species. The outcome of the response of the adults of each species to the environment and to competition from the other species is expressed simply as a ratio (seed production per plant by species A)/(seed production per plant by species B). This assumes that the ratio is unaffected by the proportions of the different species in the community. Some experiments with two species of annuals grown in different proportions do approximately conform to this (Marshall & Jain 1969; Ramakrishnan & Kumar 1971; Wu & Jain 1979).

These models predict conditions required for global stability. This means that in a given environment there is one proportion of the component species which maintains itself, and any other proportion of the species (provided they are all present) will change back to that stable proportion.

MODEL 1: TWO SPECIES, TWO DISTINCT HABITAT STATES

This model simulates a mixed-age forest containing two tree species, whose adults compete for light and perhaps mineral nutrients, without niche separation, but whose germination or seedling establishment shows niche separation. Symbols used are defined in Table 1.

TABLE 1. Symbols used in the two-species model

	Species A	Species B	Ratio A/B
Number of adult trees	n_a	n_b	
Mortality of adults per plant per year	m_a	m_b	M
Seeds produced per adult plant per year	s_a	s_b	S
Number of seeds falling in a gap	f_a	f_b	
Proportion of seeds germinating			
soil type 1	g_{1a}	g_{1b}	G_1
soil type 2	g_{2a}	g_{2b}	G_2
Number of young seedlings per gap	y_a	y_b	
Number of established saplings in forest (one per gap)	p_a	p_b	

Proportion of forest area occupied by soil type $1 = e_1$, by soil type $2 = e_2$.

The forest contains a large number of trees of species A and B. Each year a proportion of each species dies; in each resulting gap many seedlings appear, but only one survives to be a sapling which replaces the dead tree. The successful sapling may be of either species. Thus the total number of adults does not change, only the proportion of A and B. The chance that the successful colonizer of a gap will belong to species A is $y_a/(y_a + y_b)$, where y_a and y_b are the numbers of young seedlings of species A and B, respectively, in the gap. The numbers of young seedlings are determined by (1) the number of seeds of each species, f_a and f_b, falling in the gap, and (2) the proportion of these which germinate and become established. The number of seeds in each gap is proportional to the number produced in the forest as a whole, i.e. $f_a \propto n_a s_a, f_b \propto n_b s_b$; this assumes that the seeds are widely dispersed. The germination factor, g, depends on the species and the soil. There are two soil types, 1 and 2, forming patches each large enough to hold at least one tree. The number of young seedlings of species A arising in a gap on soil type 1 is given by

$$y_a = f_a g_{1a} \qquad (1)$$

with the corresponding factors g_{2a}, g_{1b}, g_{2b}. The requirement for global stability is that if either species is present in very low proportion its numbers will increase, i.e.

when $n_a \gg n_b$,

$$\frac{p_b}{p_a} > \frac{n_b m_b}{n_a m_a} \qquad (2a)$$

and when $n_b \gg n_a$,

$$\frac{p_a}{p_b} > \frac{n_a m_a}{n_b m_b} \qquad (2b)$$

where p_a and p_b are the numbers of saplings of A and B, respectively, replacing the dead trees in the gaps. Straightforward algebra (see Appendix) shows that the requirements are

$$e_1/G_1 + e_2/G_2 > S/M \qquad (3a)$$

and

$$e_1 G_1 + e_2 G_2 > M/S \qquad (3b)$$

where G_1, G_2, S and M are ratios defined in Table 1.

It is not necessary in this model to assume that the young seedlings are equally balanced in competition. If instead of $p_a = y_a/(y_a + y_b)$ we put $p_a = cy_a/(cy_a + y_b)$, where c can be above, below or equal to 1, this incorporates differences in competitive ability but mathematically the outcome is the same as if the values of g_{1a} and g_{2a} are changed.

Figure 1 shows examples of predictions from this model, taking the two soil types to occupy equal areas. The stippled portions indicate the combinations of germination and seedling survival rates of each species on each soil type which will give stability of the two species. In the remaining areas one or other species will in due course be eliminated. Figure 1(a) is for $S/M = 1$, i.e. the two species are equally balanced in competition in the adult phase. Stability can occur if species A germinates better than B on one soil type but B better than A on the other. Figure 1(a) shows how great these differences in germination need to be. In Fig. 1(b) species A is the more successful in terms of seed production and/or mortality. The shapes of the areas are, however, identical in parts (a) and (b), only the position on the graph has changed. The higher seed production or lower mortality of species A can be compensated for by lower germination in each soil type.

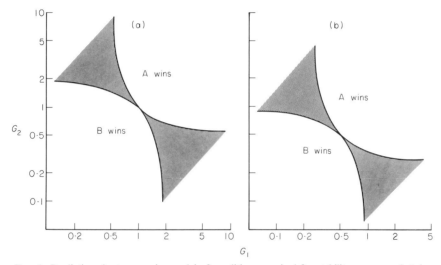

FIG. 1. Predictions by two-species model of conditions required for stability. $e_1 = e_2 = 0.5$, i.e. two soil types equal in area. Stippled portions show required values of G_1 and G_2 for stability. (a) $S/M = 1$, (b) $S/M = 2$. For definition of symbols see Table 1.

Within the stable (stippled) areas of Fig. 1, if the community is left undisturbed the proportion of A and B will reach a stable value which depends on S, M, G_1, G_2, e_1 and e_2. This stable proportion is given by

$$\frac{n_a}{n_a + n_b} = \frac{e_1(M - SG_1) + e_2(M - SG_2)}{e_1(M - SG_1)(1 - SG_2) + e_2(M - SG_2)(1 - SG_1)}. \tag{4}$$

We can use this equation to examine how the abundance of the two species will vary between communities, i.e. over a large area, in relation to change in the environment. Consider first the effect of a change in the proportion of the area occupied by the two soil types. Figure 2 shows that, as expected, an increase in the proportion of soil type 1 results in an increase of species A, whose seeds germinate better than B's on soil type 1. However, the abundance of A is not directly proportional to the abundance of soil type 1: below a certain proportion of the soil type A disappears from the community. This is because its greater success in establishing on the limited area of type 1 soil is not enough to

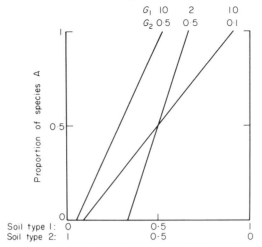

FIG. 2. Stable proportion of species A in a mixture of A and B, in relation to proportion of area occupied by the two soil types; from eqn (4). $S = 1$, $M = 1$, i.e. the two species are evenly balanced in the adult stage. For definition of symbols see Table 1.

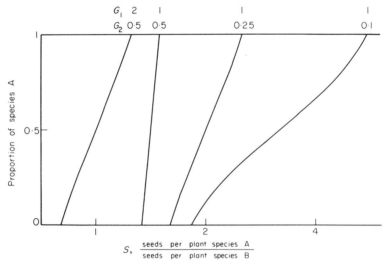

FIG. 3. Stable proportion of species A in mixture, predicted by eqn (4), in relation to seed production by the two species. $e_1 = e_2 = 0.5$, $M = 1$. For definition of symbols see Table 1.

counterbalance the success of B on the much more abundant type 2 soil. The spread of seeds from one soil type to the other is essential to this result.

Secondly, imagine an environmental gradient, e.g. of altitude, along which seed production of one species increases and the other decreases, so S changes; the proportion of the two soil types does not change. Figure 3 shows that the stable community composition would change from initially 100% of one species, through a gradually increasing proportion of the second species, until finally the first species disappeared.

MODEL 2: SIX SPECIES, CONTINUOUS HABITAT VARIATION

Six species, A–F, are all annuals and all have the same percentage mortality between germination and maturity. The mean seed production per plant for each species is $s_a, s_b \ldots$

s_f. The sizes of individual seeds of each species form a normal distribution curve with the same standard deviation, w, but different means (Fig. 4). The mean seed size of each species differs from that of the next larger or smaller species by amount d. In the soil are numerous microsites of varying sizes, an equal number in each size range, each capable of holding one seed. A seed can only germinate if it lands in a microsite of the correct size. There are enough seeds to ensure that every microsite is filled every year.

If the numbers of seeds of a particular size belonging to each species are q_a, q_b ... q_f, the chance that a seed of A occupies the site and hence germinates is $q_a/(q_a + q_b + q_c + q_d + q_e + q_f)$. q_a depends upon the total seed production by the species and upon its size distribution (see Appendix). If one specifies the values of d/w, the seed production per plant by each species and initial abundance of each species, it is possible to calculate the abundance of each species the following year. With the aid of an iterative computer program one can determine whether a particular set of d/w and seed production values leads to stable coexistence of the six species or whether one or more species become extinct. By a method of successive approximations one can determine, for any given set of seed production values, the minimum d/w, i.e. the maximum niche overlap, which permits

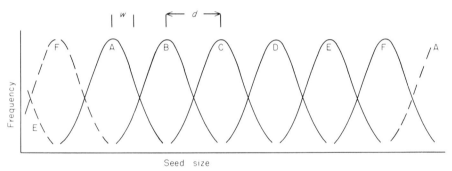

FIG. 4. Explanation of six-species model, showing size distribution of seeds of each species, A–F. Dashed lines are 'guard plants'. d = difference between means, w = standard deviation.

all the six species to coexist (see Appendix). The results tend to be substantially affected by what happens at each end of the size axis (Ågren & Fagerström 1980). For example, if available microsite sizes extend far to the left of A and to the right of F, this gives these two end species a substantial advantage. In real situations such end effects may be very important, but here I wish to avoid them, in order to obtain a more general conclusion. I have therefore used cyclic boundary conditions, creating 'guard rows' by performing the calculation as if A–F occurred again to the right and to the left of the central group, so each of the real species has at least six on each side of it in the seed size spectrum (see Appendix). There are many possible combinations of seed production per plant. I have taken as an illustration the case where the smaller the seed the greater the number of seeds produced per plant, and

$$s_a/s_b = s_b/s_c = s_c/s_d = s_d/s_e = s_e/s_f = S. \qquad (5)$$

Figure 5 shows the limiting value of d/w for a range of values of S. The result is somewhat affected if the order of the species' seed production is altered: the points marked $\times$ on Fig. 5 give some examples for $S = 2$. Intuitively one would expect that if $S = 1$, $d/w = 0$, i.e. if all species have exactly the same seed production each would maintain a constant abundance whatever their seed size distributions. In fact the computer model does not predict this; this arises because when $d/w <$ about 0·6 the 'guard plants' no longer

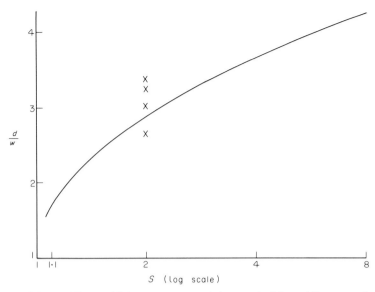

FIG. 5. Minimum difference (*d*) between mean seed sizes required for stability, as predicted by six-species model, in relation to *w*, the variability of seed size within each species and *S* the ratio between species in seed production per plant. Continuous curve: *S* as defined in eqn (5). ×: examples of results if $s_a - s_f$ rearranged within eqn (5).

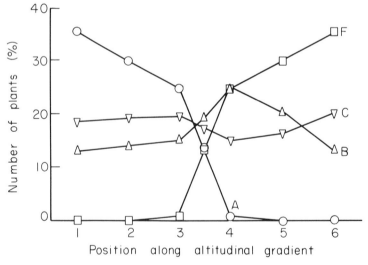

FIG. 6. Predicted abundance of each species, as percent of total, if $d/w = 2$ but seed production is varied. For clarity species D and E are omitted: their results would be the mirror images of C and B respectively. The seed production per plant assumed was:

Position on gradient	1	2	3	4	5	6
Species A	40	35	24	13	5	2
B	35	40	35	24	13	5
C	24	35	40	35	24	13
D	13	24	35	40	35	24
E	5	13	24	35	40	35
F	2	5	13	24	35	40

effectively eliminate edge effects, and the predictions for S very close to 1 are therefore of limited value. In real communities it is unlikely that seed production would be more similar than $S = 1 \cdot 1$, or more different than $S = 8$ (which would mean the largest seed production 30 000 times the smallest). Therefore the limiting value of d/w would usually be within the range approximately 2–4.

As an example of the effect of altering the number of species, with $S = 2$ and the standard arrangement of seed size, a ten-species model gave the limiting value of d/w to be $3 \cdot 4$, compared with $2 \cdot 9$ for six species. Since S was the same for both, the ratio of seed production between the highest and lowest producer was greater in the ten-species model.

As with the two-species model, one can examine what would happen over a large area within which the abundance of soil microsites remained constant but some other environmental factor, e.g. altitude, varied. Figure 6 shows an example, in which each species has been given an approximately normal distribution of seed production along an altitudinal gradient, each with its peak at a different altitude (see caption to Fig. 6). The separation of seed size between species is $d/w = 2$. Figure 6 shows that near the two ends of the gradient only five species coexist, but near the centre all six do so. These results are probably markedly influenced by details of the model. For example, the substantial changes in abundance by A and F are partly a result of having 'guard plants' in the model, making A and F compete strongly with each other for soil microsites. The main object is to illustrate how the species composition of a community could change, without any change in the abundance of the habitat states controlling niche separation. One cannot calculate d/w precisely for the curves in Fig. 6 but clearly d/w is well below 2, showing that the species-packing relationships within a community (alpha-diversity) do not necessarily apply between communities (beta-diversity). Figure 6 also shows that the order of relative abundance is not necessarily the same as the order of seed production; e.g. at point 3, C has the highest seed production but A and E are the most abundant. This is because each species is particularly affected by the species next to it in the seed size order: it competes most strongly against them for germination sites.

DISCUSSION OF THE TWO MODELS

These models, especially the first, resemble models previously described by Skellam (1951), Fagerström & Ågren (1979) and Hubbell (1980): all are concerned with regeneration from seed and in all of them the total number of adult plants is fixed but the proportion of different species can vary. However, in these previous models the basis for coexistence was different. Skellam (1951) and Fagerström & Ågren (1979) proposed that one of the two species (say B) could only colonize a gap if no seedling of the other species (A) was present. Coexistence can then occur only if A has a low seed production, usually less than about 5 per plant per year. Hubbell (1980) considered conditions under which a host-specific herbivorous insect could promote coexistence by preventing species from regenerating in their own immediate vicinity.

The two models I have proposed differ in some respects, in both their assumptions and their predictions. The key distinction is not in the number of species but in whether the habitat varies continuously. Taken together they provide some insight into how niche separation may operate in terrestrial plant communities. They show that coexistence of species is likely to occur only if each habitat state has more than a certain minimum abundance (Fig. 2). Even if the relative abundance of the different habitat states remains fixed the number of species coexisting and their relative abundance can change (Figs 3 and 6). This comes about by a change in the competitive balance between adult plants altering their relative seed production. Figures 3 and 6 show that as some environmental factor changes, for example moving along a large-scale gradient of temperature, moisture or soil fertility, a species at first absent can gradually increase in abundance until it is the most abundant species. Thus gradual changes in abundance seen in real vegetation similar to those of Fig. 6 (e.g. Whittaker 1956), are in no way incompatible with niche separation, and there is no need to postulate that the abundance of different habitat states is changing along the environmental gradient. The models predict that in a particular stand the abundance of the different species is not necessarily directly proportional to either the abundance of the habitat states they favour or their seed production.

The models can be extended to cover many situations where regeneration from seed might provide niche separation. The six-species model was presented as applying to annuals, but it could apply equally well to perennials, provided they all had the same average life-span. It could, for example, be applied to year-to-year fluctuations, the horizontal axis of Fig. 4 being some weather factor, and the vertical axis seed production, germination or seedling survival. If varying seed production is to provide a mechanism for coexistence, most seed must germinate in the first year after production, and the plant's average life-span must be long relative to weather fluctuations. If year-to-year variations in germination or seedling survival are involved, then either the plants or the dormant seed pool must be long-lived.

Model 1 predicts that when there are two distinct habitat states, two species can coexist at any combination of G_1 and G_2, provided S/M is right. However, some combinations of G_1 and G_2 give stability only over a very limited range of S/M (Figs 1 and 3), which would make permanent coexistence unlikely in reality. If $G_1/G_2 > 4$ or $G_2/G_1 > 4$ the range of S/M for stability is more extensive: this defines the squares in the bottom right and top left of Fig. 1(a) and 1(b), bounded by asymptotes.

The prediction of the six-species model, that when the habitat varies in a continuous manner niche separation needs to be $d/w > 2$ to 4 for stability, may be compared with the requirement of $d/w > 1.6$, 1, 1.4 previously predicted by MacArthur (1972), May (1974) and Ågren & Fagerström (1980) respectively. My model thus appears to require greater separation. However, in the MacArthur and May models substantial differences between species in 'carrying capacity' can raise the required d/w (e.g. May 1974, Fig. 6.6). The carrying capacity is in some ways analogous to the seed production in my models, and the differences in prediction may therefore be less than at first appears.

Sometimes d/w cannot be calculated from real data, but an approximate test can still be applied. For example, seed production in several years may be known but the cause of the variation be uncertain. If the seed production in each year is expressed as per cent frequency, i.e. as a percentage of the total production by that species in all recorded years, and if $d/w = 2$, then at its peak (modal) seed production a species will have $7.4 \times$ the seed production of the next highest species, and half a standard deviation to right or left it will have $2.7 \times$ the next highest. Therefore if n species are to coexist, at least once in every $2n$ yr (on average) each species should have a year in which its seed production is about three times or more higher than any other species.

APPLICATION TO REAL DATA

Separation in space

King (1977a) showed that in an English chalk grassland some species were more abundant on ant-hills than in the surrounding sward while for other species the reverse was true. Sometimes this could be attributed to differences in germination (King 1975, 1977b). For example, *Arenaria serpyllifolia** was commoner on ant-hills, but *Leontodon hispidus* in the sward. Their germination on ant-hills was *A. serpyllifolia* 41%, *L. hispidus* 4·8% (ratio, $G_1 = 8·5$); in the sward *A. serpyllifolia* had 4·8% germination, *L. hispidus* 8·4% ($G_2 = 0·6$). Here the requirement $G_1/G_2 > 4$ is met, with plenty to spare.

Miles (1974) cut bare areas 25, 250 or 2500 cm² in area in *Calluna vulgaris* heath in Scotland, simulating natural disturbance. He recorded seedlings which established in each of the following two years; Fig. 7 shows the results, expressed as frequency, for some of

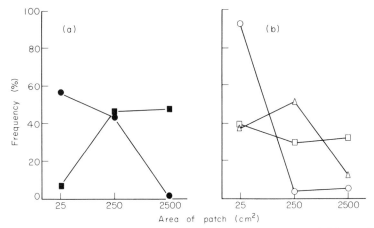

FIG. 7. Seedling emergence, as percentage of total for each species, in gaps of different sizes in Callunetum. (a) Shrubs: ●, *Calluna vulgaris*; ■, *Cytisus scoparius*. (b) Herbs: ○, *Galium verum*; □, *Lotus corniculatus*; △, *Thymus praecox*. Data of Miles (1974).

the species. Although patch size is essentially a continuous variable, the responses of species to it do not form normal distribution curves, and values of d and w cannot be calculated. However, by the requirement that each species has a patch size in which it does at least three times as well as any other species, the two shrub species present show fully adequate niche separation (Fig. 7(a)). Each of the three herbaceous species shown in Fig. 7(b) does better than the other two at one patch size, but *T. praecox* never approaches three times the value of the other two. Therefore, by the model's requirements, *G. verum* could coexist with either of the other two species, but not all three together. Of course a larger range of patch sizes might provide further variation. The other herbaceous species which germinated had responses similar to, or intermediate between, the three shown in Fig. 7(b).

Harper, Williams & Sagar (1965) sowed seeds of three *Plantago* species uniformly over a bed of soil to which various treatments were then applied to produce local microhabitats;

* Nomenclature of British species follows Clapham, Tutin & Warburg (1981).

TABLE 2. Number of seedlings of three species of *Plantago* which appeared per dm²
in different areas of plots in experiment of Harper, Williams & Sagar (1965)

	Treatment*			
	Hole	Wall	Sheet	Control
P. lanceolata	13·2	6·0	5·4	0·59
P. major	4·2	1·1	2·6	0·79
P. media	0·98	0·33	4·0	0·11
Ratio:				
lanceolata/major	3·2	5·6	2·1	0·74
lanceolata/media	14	18	1·3	5·2
major/media	4·3	3·3	0·65	7·0

* Treatments: hole 12·5 cm × 12·5 cm × 2·5 cm deep; wooden walls 2·5 cm high; glass
sheet flat on soil surface; untreated soil surface.

e.g. holes were dug or vertical barriers erected simulating such natural features as cattle
hoof-prints and dead grass tussocks. Table 2 shows some of their results. All three species
germinated better near the treated areas than on the level, untreated soil surface, but there
were differences which could provide adequate niche separation, by the rule that G_1/G_2
must exceed 4. Because *P. lanceolata* showed a much larger difference than *P. major*
between treated and untreated areas, a mixture of holes and untreated or walls and
untreated gives $G_1/G_2 > 4$ for these two species. For a combination of *P. media* with either
of the others, $G_1/G_2 > 4$ can be achieved only if one of the microsites is a sheet of glass laid
on the soil surface. I cannot think what natural microhabitat this simulates.

Oomes & Elberse (1976) performed rather similar experiments in which seeds of six
grassland species were sown on bare, level soil and in grooves of several widths. Again
species usually germinated better in grooves than on the level soil surface, but the
differences in response were large enough to give $G_1/G_2 > 4$ for some species
combinations. The responses to different groove widths were not sufficiently different to
provide any further niche separation.

Although it is often suggested that germination is controlled by 'safe sites' each holding
a single seed, I know of no data showing the relationship of germination to the size or
shape of the microsite within which the individual seed is located. If niche separation is by
the size of the microsites a minimum requirement would be that seed size differences
between species give $d/w > 2$. Four winter annual species (Table 3) grow on shallow soils
over limestone on the Mendip Hills near Bristol. They are apparently very similar in the
timing of their life cycle. All have ovoid seeds, and it is interesting to know whether the
seeds differ sufficiently in size to make this a possible mechanism of niche separation.
Seeds collected by V. Cowling from Ubley Warren, Somerset, were measured with a

TABLE 3. Mean and standard deviation (*μ*m) of length and breadth of seeds of four
winter annual species. Each figure is based on measurements of fifty seeds. A few
seeds which were clearly deformed were excluded. Measurements by S. Graves

	Length			Breadth		
	Mean	Difference	Standard deviation	Mean	Difference	Standard deviation
Cardamine hirsuta	914		59	706		88
Hornungia petraea	778	136	28	538	168	38
Erophila verna	586	192	48	382	156	35
Cerastium pumilum	507	79	60	404	−22	65
Mean of s.d.		49				57

micrometer eyepiece. Table 3 shows that three of the species can be separated from each other by $d/w > 2$, in both length and breadth, but *E. verna* is by this rule not sufficiently different from *C. pumilum* to provide a mechanism for coexistence.

Separation in time

In deserts where rainfall is only occasionally sufficient to allow much germination, the temperature at the time of rainfall can determine the proportions of different species which germinate from the seed pool (Went 1948, 1949). When rain is commonest at two seasons this can result in two distinct populations of ephemerals within a year (Went 1948; Mott 1972). One may ask whether the temperature responses of the species differ enough to allow niche separation between years, depending on the temperature when rain falls, or only between two seasons within the year. Went (1949) determined the amount of germination of ephemerals from Californian deserts at various temperatures, and Mott (1972) obtained similar data for three desert ephemerals of Western Australia. The results are summarized in Table 4. The winter- and summer-germinators are approximately

TABLE 4. Relationship of germination to temperature in desert ephemerals. Modes and standard deviations determined from smooth curves drawn through data points by eye. Data from: (a) Went (1949), (b) Mott (1972)

	Modal temperature (°C)	Standard deviation (°C)
(a) Species from Californian deserts		
Pectis papposa	27	3
Amaranthus deflexus	25	2
Bouteloua barbata	23	4
Filago arizonica	14	6
(b) Species from W. Australian desert		
Aristida contorta	29	10
Helichrysum cassinianum	17	7
Helipterum craspedioides	14	8

separated by $d/w = 2$, but within each of those two groups the separation is much less, indicating no opportunity for niche separation between years.

In moister climates species can also differ in their season of germination. Ellenberg (1963) quotes observations of Salzmann on the germination during each month from November to June of twenty-eight species of weed in fields in Europe. Although d and w cannot be calculated precisely, one can pick from the list four species which show so little overlap in germination period that clearly $d/w > 2$; each has more than 90% of its total germination in the period November, December, March, April–June, respectively.

Davies (1976) recorded the fruit production of ten species of trees and shrubs in 9 yr in an arid area of Western Australia. All but one showed marked variation in production from year to year. Fruit production was related to weather in a complex fashion which precludes application of the d/w rule. If all ten species are to coexist by the variation in seed production, then by the 'once in $2n$ years rule', we should expect that in 9 yr there would be at least 4 or 5 yr in which one species, a different one each time, had its fruit production, expressed as a percentage of its total, at least three times as high as any other species. In fact in Davies' data this was not true for any year. It is possible to pick out a group of three species each of which is at least three times higher than the other two in a particular year, or indeed two such groups, but not a group of four. So Davies' data suggest that weather fluctuations might account for sets of three species coexisting but not more.

CONCLUSIONS

These examples show the value of having several alternative niche separation rules which can apply to different sorts of data. In none of the examples given here did application of the tests provide positive evidence that species would form a stable mixture: to provide that, more information would be needed, e.g. about mortality in the vegetative phase. The value of the tests is negative, to indicate when coexistence would probably *not* be permitted solely by the niches considered, or to indicate the *maximum* number of species which could be permitted to coexist. The examples cited, though limited in number, suggest that usually two or three species is the most which can achieve adequate niche separation along one habitat axis. This implies that if the coexistence of many species is to be explained by fixed habitat states then several different 'niche dimensions' must be involved. For example, in the heathland studied by Miles (1974), the two shrubs may be separated in 'niche space' from the herbs, e.g. by having fatter roots which favour larger soil pores. The response of seeds to patch size could then provide niche separation between the two shrubs and also two herb species. In theory six niche axes with room for only two species along each dimension could allow coexistence of $2^6 = 64$ species provided that the axes were independent, i.e. if all the sixty-four combinations of habitat states actually existed. Whether this happens in reality remains to be demonstrated.

These examples provided no test of the accuracy or validity of the models' predictions. The models involve various simplifications and assumptions, and none of the examples conformed exactly to these. Until independent tests of predictions from the models have been made any conclusions from them must be taken with great caution. Nevertheless, the models help to illustrate the part which fixed habitat states play in the maintenance and control of species diversity. Although diversity is likely to be influenced by the heterogeneity of the habitat, the models show how the number of species in a community and their relative abundance need not be precisely related to the abundance of the habitat states which favour each species, but can be markedly influenced by competition for other factors unrelated to niche separation. Therefore marked changes in abundance of individual species, or even in species-richness, occurring over time or space, could occur without change in the habitat states providing the niche separation (see Figs 3 and 6) and are not inconsistent with niche separation being provided by fixed habitat states.

The emphasis in this paper on regeneration niches does not imply that I consider these especially important, only that this is one type of niche separation worthy of study. Indeed, application of these untested models to these few examples suggests that regeneration niches with fixed habitat states are on their own not enough to account for the diversity of most communities, though they could make a contribution.

ACKNOWLEDGMENTS

I thank G. I. Ågren, T. Fagerström, P. J. Grubb and D. McC. Newbery for helpful comments on an earlier draft of this paper.

REFERENCES

Ågren, G. I. & Fagerström, T. (1980). On environmental variability and limits to similarity. *Journal of Theoretical Biology*, **82**, 401–404.

Clapham, A. R., Tutin, T. G. & Warburg, E. F. (1981). *Excursion Flora of the British Isles*, 3rd edn. Cambridge University Press.

Davies, S. J. J. F. (1976). Studies of the flowering season and fruit production of some arid zone shrubs and trees in Western Australia. *Journal of Ecology*, **64**, 665–687.

Ellenberg, H. (1963). *Vegetation Mitteleuropas mit den Alpen*. Eugen Ulmer, Stuttgart.

Fagerström, T. & Ågren, G. I. (1979). Theory for coexistence of species differing in regeneration properties. *Oikos*, **33**, 1–10.

Grubb, P. J. (1977). The maintenance of species-richness in plant communities: the importance of the regeneration niche. *Biological Reviews*, **52**, 107–145.

Harper, J. L., Williams, J. T. & Sagar, G. R. (1965). The behaviour of seeds in soil. I. The heterogeneity of soil surfaces and its role in determining the establishment of plants from seed. *Journal of Ecology*, **53**, 273–286.

Hubbell, S. P. (1980). Seed predation and the coexistence of tree species in tropical forests. *Oikos*, **35**, 214–229.

King, T. J. (1975). Inhibition of seed germination under leaf canopies in *Arenaria serpyllifolia*, *Veronica arvensis* and *Cerastium holosteoides*. *New Phytologist*, **75**, 87–90.

King, T. J. (1977a). The plant ecology of ant-hills in calcareous grasslands. I. Patterns of species in relation to ant-hills in southern England. *Journal of Ecology*, **65**, 235–256.

King, T. J. (1977b). The plant ecology of ant-hills in calcareous grasslands. III. Factors affecting the population sizes of selected species. *Journal of Ecology*, **65**, 279–315.

MacArthur, R. H. (1972). *Geographical Ecology*. Harper & Row, New York.

MacArthur, R. H. & Levins, R. (1967). The limiting similarity, convergence, and divergence of coexisting species. *American Naturalist*, **101**, 377–385.

Marshall, D. R. & Jain, S. K. (1969). Interference in pure and mixed populations of *Avena fatua* and *A. barbata*. *Journal of Ecology*, **57**, 251–270.

May, R. M. (1974). *Stability and Complexity in Model Ecosystems*, 2nd edn. Princeton University Press.

Miles, J. (1974). Effects of experimental interference with stand structure on establishment of seedlings in Callunetum. *Journal of Ecology*, **62**, 675–687.

Mott, J. J. (1972). Germination studies on some annual species from an arid region of Western Australia. *Journal of Ecology*, **60**, 293–304.

Oomes, M. J. M. & Elberse, W. T. (1976). Germination of six grassland herbs in microsites with different water contents. *Journal of Ecology*, **64**, 745–755.

Platt, W. J. & Weis, I. M. (1977). Resource partitioning and competition within a guild of fugitive prairie plants. *American Naturalist*, **111**, 479–513.

Ramakrishnan, P. S. & Kamur, S. (1971). Productivity and plasticity of wheat and *Cynodon dactylon* (L.) Pers. in pure and mixed stands. *Journal of Applied Ecology*, **8**, 85–98.

Rappoldt, C. & Hogeweg, P. (1980). Niche packing and number of species. *American Naturalist*, **116**, 480–492.

Rogers, R. W. & Westman, W. E. (1979). Niche differentiation and maintenance of genetic identity in cohabiting *Eucalyptus* species. *Australian Journal of Ecology*, **4**, 429–439.

Sheikh, K. H. & Rutter, A. J. (1969). The responses of *Molinia caerulea* and *Erica tetralix* to soil aeration and related factors. I. Root distribution in relation to soil porosity. *Journal of Ecology*, **57**, 713–726.

Skellam, J. G. (1951). Random dispersal in theoretical populations. *Biometrika*, **38**, 196–218.

Watt, A. S. (1947). Pattern and process in the plant community. *Journal of Ecology*, **35**, 1–22.

Weaver, J. E. (1919). *The Ecological Relations of Roots*. Carnegie Institution of Washington Publication no. 286.

Went, F. W. (1948). Ecology of desert plants. I. Observations on germination in the Joshua Tree National Monument, California. *Ecology*, **29**, 242–253.

Went, F. W. (1949). Ecology of desert plants. II. The effect of rain and temperature on germination and growth. *Ecology*, **30**, 1–13.

Whittaker, R. H. (1956). Vegetation of the Great Smoky Mountains. *Ecological Monographs*, **26**, 1–80.

Wu, K. K. & Jain, S. K. (1979). Population regulation in *Bromus rubens* and *B. mollis*: life-cycle components and competition. *Oecologia*, **39**, 337–357.

Yoshiyama, R. M. & Roughgarden, J. (1977). Species packing in two dimensions. *American Naturalist*, **111**, 107–121.

APPENDIX

Model 1

Let the total number of trees be T. All other symbols are defined in Table 1.

In each year $n_a m_a$ trees of species A die, and $n_b m_b$ of species B die. The resulting gaps are colonized by p_a and p_b saplings of species A and B respectively. Therefore whether the proportion of A in the forest increases, decreases or remains the same depends on whether p_a/p_b is more than, less than or equal to $n_a m_a/n_b m_b$.

Consider a single gap on soil type 1. The number of seeds falling on it, of species A and B respectively, is $n_a s_a/T$ and $n_b s_b/T$. The numbers which germinate, y_a and y_b, are given by

$$y_a = \frac{g_{1a}\, n_a s_a}{T}$$

$$y_b = \frac{g_{1b}\, n_b\, s_b}{T}.$$

The probability that the new sapling in the gap is of species A is

$$\frac{y_a}{y_a + y_b} = \frac{g_{1a}\, n_a\, s_a}{g_{1a}\, n_a\, s_a + g_{1b}\, n_b\, s_b}.$$

Since there are $e_1(n_a m_a + n_b m_b)$ gaps on soil type 1, the total number of new saplings of species A on soil type 1 is

$$e_1(n_a m_a + n_b m_b)\left(\frac{g_{1a}\, n_a\, s_a}{g_{1a}\, n_a\, s_a + g_{1b}\, n_b\, s_b}\right).$$

Adding the number of new saplings on soil type 2,

$$p_a = (n_a m_a + n_b m_b)\left(\frac{e_1\, g_{1a}\, n_a\, s_a}{g_{1a}\, n_a\, s_a + g_{1b}\, n_b\, s_b} + \frac{e_2\, g_{2a}\, n_a\, s_a}{g_{2a}\, n_a\, s_a + g_{2b}\, n_b\, s_b}\right)$$

$$\therefore\ \frac{p_a}{p_b} = \frac{n_a\, s_a \left(\dfrac{e_1\, g_{1a}}{g_{1a}\, n_a\, s_a + g_{1b}\, n_b\, s_b} + \dfrac{e_2\, g_{2a}}{g_{2a}\, n_a\, s_a + g_{2b}\, n_b\, s_b}\right)}{n_b\, s_b \left(\dfrac{e_1\, g_{1b}}{g_{1a}\, n_a\, s_a + g_{1b}\, n_b\, s_b} + \dfrac{e_2\, g_{2b}}{g_{2a}\, n_a\, s_a + g_{2b}\, n_b\, s_b}\right)}. \tag{A1}$$

To find the stable proportions of A and B, set the right-hand side of this equation equal to $n_a m_a/n_b m_b$. By rearranging and substituting in M, S, G_1 and G_2, this leads directly to eqn (4) in the text. To derive inequality (3a), set $n_a \gg n_b$. As $n_b \to 0$, the right-hand side of eqn (A1) tends to

$$\frac{e_1 + e_2}{\dfrac{n_b\, s_b}{n_a\, s_a}\left(\dfrac{e_1\, g_{1b}}{g_{1a}} + \dfrac{e_2\, g_{2b}}{g_{2a}}\right)} = \frac{n_a\, s_a}{n_b\, s_b}\left(\frac{1}{e_1/G_1 + e_2/G_2}\right).$$

The community is stable if

$$\frac{n_b\,m_b}{n_a\,m_a} < \frac{p_b}{p_a},$$

i.e. if

$$\frac{n_b\,m_b}{n_a\,m_a} < \frac{n_b\,s_b}{n_a\,s_a}\left(\frac{e_1}{G_1} + \frac{e_2}{G_2}\right).$$

Substituting in S and M leads directly to inequality (3a). Derivation of (3b) is strictly analogous.

Model 2

Let species A have mean seed size $\bar{x}_a$ with standard deviation w; let there be n_a plants of species A with mean seed production per plant s_a. Consider a particular seed size, x, and ignore for the moment the 'guard plants'. The total number of seeds of species A of this size, q_a, is given by $n_a s_a f_{xa}$, where f_{xa}, the proportion of seeds of species A of this size, is given by the equation for the normal distribution curve

$$f_{xa} = \frac{1}{w\sqrt{2\pi}} \cdot e^{-(x-\bar{x}_a)^2/2w^2}.$$

Now introducing the 'guard plants': their mean seed sizes are $\bar{x}_a + 6d$ and $\bar{x}_a - 6d$, and their standard deviations are w. Therefore

$$q_a = \frac{n_a s_a}{w\sqrt{2\pi}} \cdot \left(e^{-(x-\bar{x}_a)^2/2w^2} + e^{-(x-\bar{x}_a-6d)^2/2w^2} + e^{-(x-\bar{x}_a+6d)^2/2w^2}\right).$$

In the computer program the x axis was 120 units long, and the mean seed sizes of A–F were fixed at 10, 30, 50, 70, 90, 110, so d was fixed at 20. The values of $s_a \ldots s_f$ and w and the initial values of $n_a \ldots n_f$ were specified. Then the computer worked along the x axis one unit at a time calculating q for each species at each point, and hence the number of seeds of each species of each size which would germinate. By summing across all sizes the number of plants of each species in the next generation was determined. Using an interactive computer one can follow the changes in proportions of the species over many generations until a stable mixture is reached. The minimum value of d/w allowing all six species to coexist was found by using the computer to determine the proportion of each species in stable six-species mixtures. Using values of w which gave stable proportions of the least abundant species close to 0, graphical extrapolation could determine accurately the value of w at which this species reached 0, this being the largest value of w at which the six species coexist.

The program was written in Multics Basic.

THE CONTROL OF RELATIVE ABUNDANCE IN COMMUNITIES OF HERBACEOUS PLANTS

P. J. GRUBB, D. KELLY AND J. MITCHLEY

Botany School, University of Cambridge, Cambridge CB2 3EA, U.K.

SUMMARY

It is usual to find that the abundance of one species relative to that of another varies appreciably from year to year, and from place to place, within a single plant community. It is also common to find that certain species are persistently more abundant than others, and in some communities quite long and relatively fixed hierarchies are found.

In order to interpret changes in relative abundance one must consider the interaction of all stages of the life-cycle with changes in the weather, and variations in soil-surface characteristics and incidence of predators. Where one species is persistently less abundant than another, it is because either there are fewer micro-sites in which it can regenerate, or it is less effective in reaching the micro-sites suitable for it. Less abundant species can persist in a community where they find fewer appropriate micro-sites provided they are quicker to invade them, or have a greater potential for interference in the short term. Less abundant species can persist where they are less effective in reaching suitable micro-sites provided they have a greater potential for interference in those sites.

INTRODUCTION

Much has been written in recent years about the factors controlling the abundance of single plant species, and a masterly summary has been provided by Harper (1977). In contrast, much less attention has been paid to the problem of why one species is more common than another, especially in natural and semi-natural vegetation. In this article we discuss the control of relative abundance in herbaceous communities. By 'relative abundance' we mean simply the abundance of one species relative to that of another at a given place and time. It is usual to find appreciable variation in relative abundance from time to time and place to place even in what is commonly regarded as a single 'community' occupying a relatively uniform habitat. However, it is also often found that in a particular community certain species are regularly more abundant than others. Indeed quite long and relatively fixed hierarchies may be found. A. S. Watt had these in mind when he wrote (1961, p. 125) 'as far as I am aware, no critical examination has been made to find out why, in a given plant community some species are rare, some common, some constant (even if rare or occasional only) and others sporadic.' It appears that during the last 20 yr few ecologists have taken up the challenge, and attempted the necessary 'critical examination'.

Our attempt at such an examination is set in the broader context of considering the control of relative abundance generally, including those situations where at first sight the elements of variation are more apparent than the elements of constancy. Abundance in the broadest sense may be analysed into two rather different properties: frequency in samples

0262–7027/82/0300–0079$02.00 © 1982 British Ecological Society

of a given size, and what we may call 'abundance where found'. We shall be chiefly concerned with the latter, although spatial distribution will certainly be discussed. Depending on context, 'abundance' may be measured as numbers of plants, cover or above-ground biomass. We deliberately use the neutral term relative abundance rather than 'dominance', which has come to mean different things to different people. We are most often concerned with relative abundance within one life-form (relatively widely defined) but in some cases the term is used in comparing appreciably different life-forms, e.g. long-lived tussock-forming grasses and annual or biennial dicotyledons.

In the first half of the paper we define our problem more closely, using examples from various parts of the world. In the second half we put forward a tentative basis for understanding, and emphasize the huge gaps in our present knowledge. Our nomenclature follows Flora Europaea (Tutin *et al.* 1964–80), except where we cite papers describing work done outside Europe and adopt the nomenclature used in those papers.

DEFINITION OF THE PROBLEM

Range of communities considered

We consider communities of three types: (i) communities composed entirely of annuals, (ii) communities composed of species of various life-lengths and subject to continual disturbance, and (iii) similar communities subject to periodic rather than continual disturbance. Following Grime (1979, p. 39), we define disturbance as consisting of the 'mechanisms which limit plant biomass by causing its partial or total destruction', but we add the rider that the mechanisms must be generated outside the community and do not include senescence of plants.

Communities of annuals

We review firstly those communities where 'interference', i.e. infliction of hardships on neighbours (Harper 1961), is manifestly important, and secondly those communities where interference is not obvious. In the first case we have dense natural communities of annuals in many semi-deserts, and on coast drift-lines and river banks in the temperate zones, as well as man-made communities of crops and their weeds. In the second case we have very sparse semi-natural communities such as those composed of winter annuals at moderately disturbed sites on sand dunes in the temperate zones, where temperatures and irradiance levels are low during the growing season, the supplies of water and mineral nutrients are deficient, and the plants are subject both to erosion of sand around their roots and deposition of sand over their shoots (Pemadasa & Lovell 1976).

Wherever studies of abundance have been made over extensive areas and a number of years in communities of annuals showing manifest interference, great variation has been found in both total cover and spatial distribution of most species. We use some data of Torssell (1973) to illustrate this point. His study was made in a monsoonal climate in the Northern Territory of Australia. On many beef-cattle properties there, the native savanna woodland has been cleared and replaced by mixtures of the introduced annual C_3 legume *Stylosanthes humilis* and indigenous annual C_4 grasses, chiefly *Brachiaria ramosa* and *Digitaria ciliaris*. The large year-to-year changes in absolute abundances, relative abundances and spatial distributions are clearly seen in Fig. 1.

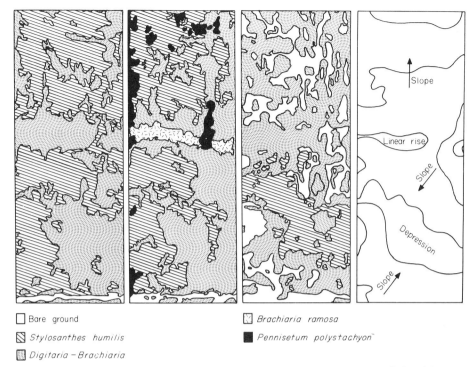

☐ Bare ground

▨ Stylosanthes humilis

▦ Digitaria – Brachiaria

☒ Brachiaria ramosa

■ Pennisetum polystachyon‾

FIG. 1. Areas in which various species were most abundant in an annual pasture 2, 4 and 6 yr after sowing *Stylosanthes* on to newly ploughed land on a plot 180 × 70 m in the Northern Territory of Australia (after Torssell 1973); the drawings are based on photographs taken at a height of 300 m.

A further point to be made in Fig. 1 is the fact that certain grasses (the *Brachiaria* and *Digitaria*) are consistently more abundant than others such as *Echinochloa crus-galli* and *Pennisetum polystachyon*. More details of the relative abundances of these grasses are provided by Torssell (1973) in his Table 1. An analagous community has been well documented in California, in regions where European annual weeds have largely displaced native species in grazed woodland. Very commonly a few species make up most of the plant cover, and there is a long list of rare species. Thus Talbot, Biswell & Hormay (1939) found that in an area of 1440 acres (583 ha) with 109 species of annual, three species were consistently the most abundant over a period of 3 yr. A summary of their data is given in our Table 1. Their two most abundant species, *Bromus hordaceus* and *Erodium botrys*, have been found to be the most abundant annuals year after year in many other parts of California (McCown & Williams 1968).

TABLE 1. The percentage composition of herbaceous cover in six pastures on the San Joaquin Experimental Range, California in three successive years, based on 1440 random samples of 1 ft² (0·093 m²) taken 1 per acre (0·4 ha) in 1936–7, and 720 samples (1 per 2 acre) in 1938 (from Talbot *et al.* 1939)

	1936	1937	1938
Erodium botrys	40·6	34·3	15·8
Bromus hordaceus	19·4	30·6	17·2
Festuca megaleura	16·9	9·3	10·3
Next most abundant species	2·8	4·2	5·1

Our two examples of communities with manifest interference thus illustrate well the twin problems that have to be answered in studying the control of relative abundance: how is the variation from time to time and place to place brought about, and how is it that some species can be consistently more abundant than others without the less abundant species being lost completely? The same problems have emerged from a study made by one of us (D.K.) on a sparse community of annuals in which interference is not obvious. The community studied consists of winter annuals on fixed dunes in a nature reserve on the coast of eastern England. The plants concerned flower and fruit when they are 1–2 cm tall, and the shoot of each occupies an area of 0·5–3 cm²; the plants commonly occur at densities of 100–1000 m⁻², and cover only 1–10% of the ground. The cover of grasses and sedges is less than 10%, and there is a variable cover of moss. The community is kept open by rabbit grazing, human trampling and slight accretions of blown sand. The numbers of flowering individuals have been recorded in 500 quadrats of 100 cm² in each of two fixed plots of 5 m² over 3 yr. The original data show much evidence of year-to-year change in the areas of maximal abundance of each species. However, as shown in Table 2, there is also evidence of a consistent hierarchy of abundance over the whole 5 m² in the area where

TABLE 2. Numbers of flowering individuals of four winter annuals in two fixed plots of 5 m² on sand dunes on the north Norfolk coast in three successive years*

	Area X (moss covering increasing 1979–1981)			Area Y (bare sand increasing 1980–1981)		
	1979	1980	1981	1979	1980	1981
Erophila verna	813	1196	3496	402	444	767
Cerastium semidecandrum	75	473	1353	1452	2351	718
Myosotis ramosissima	16	78	380	422	647	661
Valerianella locusta	1	23	45	224	619	478

* Records made on Holkham National Nature Reserve (National Grid reference TF 860457).

the moss cover has been increasing. Preliminary observations on comparable sites with low moss cover at other localities within 20 km of the one studied in detail suggest that it is usual for the *Cerastium* or *Erophila* to be the most abundant species, the *Myosotis* intermediate, and the *Valerianella* least common.

Communities composed of plants of various life-lengths and subject to continual disturbance

We have in mind chiefly grazed and mown communities. Such communities suffer not only continual major disturbance through removal of shoot material, but also periodic minor disturbance from heaping of soil by moles, earthworms and ants, scraping by rabbits, and trampling by animal feet or compression by tractor wheels.

Several studies have emphasized the variation in relative abundance from year to year in such communities. For example, Rabotnov (1966) made this point for mown grasslands in the Oka Valley in Russia. He presented data for yields of fifteen species on a plot of 8 m² over 10 yr (his Table 1). We give in Table 3 the ranks of the fifteen species in the 10 yr. The errors in harvesting and weighing were not reported, and arguably more ranks should be presented as ties. Three points of general interest may be made on the basis of Table 3. First, if a biologically significant change in rank is deemed to be one that involves a change

Table 3. The ranks in 1954–63 of the fifteen species in an 8-m² sample of a valley meadow in Russia contributing 5% or more to the aboveground biomass in at least one year, based on data of Rabotnov (1966). The vertical bars indicate change of four or more in rank. The total yield (kg per 8 m²) is also shown

	54	55	56	57	58	59	60	61	62	63	Mean rank
Festuca pratensis	6	4	7	5	5	2	5	3	2	1	4·0
Poa pratensis	7	8	9	10	1	1	2	1	1	2	4·2
Lathyrus pratensis	10	9	4	3	2	3	1	5	5	5	4·7
Cirsium arvense	4	1	1	1	3	8	6	9	7	8	4·8
Achillea millefolium	9	6	3	4	4	11	10	4	3	6	6·0
Bromus inermis	5	5	5	8	9	4½	7	8	12	11	7·5
Vicia cracca	11	10	6	6	11	10	9	11	6	4	8·4
Campanula glomerata	15	14	14	12	6	7	3	2	4	7	8·4
Seseli libanotis	12	11	11	9	7½	9	4	6	10	9	8·9
Phleum pratense	2	3	12	13	7½	6	13	10	14	14	9·4
Trifolium pratense	3	12	2	2	14	14	14	12	11	10	9·4
Elymus repens	8	7	10	11	10	4½	11	13	9	12	9·6
Agrostis stolonifera	1	2	13	15	12	13	12	14	13	3	9·8
Silene vulgaris	14	13	15	14	15	12	8	7	8	13	11·9
Trifolium hybridum	13	15	8	7	13	15	15	15	15	15	13·1
Total yield	1·9	2·3	1·5	2·4	2·4	1·9	1·0	1·8	1·5	1·6	

of at least four places, then in one year there was no significant change (1956–57), and in another year only one (1954–55), whereas in other years 5–6 species changed rank significantly (1955–56, 1957–58, 1959–60, 1961–62). Secondly, the major changes in different species were staggered; for example, the biggest change in 10 yr for *Poa pratensis* was in 1957–58, and that for *Cirsium arvense* in 1958–59, while the two major changes for *Lathyrus pratensis* were in 1955–56 and 1960–61. Thirdly, species that have coupled changes in one year show uncoupled changes in other years; *Phleum pratense* and *Trifolium pratense* illustrate this point well.

If Friedman's Test for Randomized Blocks (Sokal & Rohlf 1969, p. 397) is applied to the rank orders in Table 3, there is found to be a significant constancy of rank ($P <$ 0·001). This statistical result arises largely from the relatively constant positions of several species over runs of 5–8 yr. It cannot blind us to the fact that the hierarchy was very different in, say, 1954 and 1963. If Kendall's Rank Correlation Test is applied to the ranks for these 2 yr, there is no significant correlation ($P = 0·29$).

Unfortunately we have no evidence about the extent to which the changes recorded were typical of the whole meadow, let alone the meadows in general in the Oka Valley. It may well be that over a large area there was no change in 1954–62 from *Agrostis* and *Phleum* to *Poa* and *Festuca* as the leading grasses. Possibly there was only a shuffling of the areas occupied by these plants.

A contrast to Rabotnov's river meadow is provided by the *Festuca duriuscula-F. psammophila* grassland on an inland fixed dune system in Poland studied by Symonides (1979). Over the period 1968–75 the relative abundance of most of the thirty-six species recorded in 256 m² changed remarkably little.

In the last few years we have made a detailed study of another community-type which shows little year-to-year variation in the relative abundances of the perennials. This is 'chalk grassland', i.e. grassland on Cretaceous limestones: the soils are relatively infertile, and traditionally the grasslands have been grazed rather than mown. We have taken the precaution of making observations at two widely separated sites (in Sussex and Wiltshire, 150 km apart), and we are able to compare our data with those obtained by another ecologist at a series of sites 12–15 yr earlier. At both our study sites the turf has generally been kept to <8 cm (excluding inflorescences) in recent years. The Sussex plot is steeper (15–20° *v.* 5–10°), has more bare ground (5% *v.* 0·4% in August 1980), and is grazed only from September to March whereas the Wiltshire site is grazed all the year. The plots are on south- and south-west-facing slopes respectively.

At each site a plot of 80 × 48 m has been chosen, and divided into twenty subplots of 16 × 12 m. In each subplot one stand of 0·48 m² has been chosen at random, and marked permanently. One of us (J.M.) has then recorded the plants intercepted by a fine sewing needle at 120 or 240 randomly chosen points in each stand (2400 or 4800 in all) three times a year for 2 yr (early summer, late summer and autumn).

We present data here for non-repetitive cover for only the longer-lived plants. The arbitrarily separated shorter-lived plants are dealt with below, using a different technique. We consider first the Sussex data for August 1980. Of the thirty-five perennials concerned 40% were intercepted in all twenty stands of 0·48 m², and 74% in at least sixteen out of twenty. Clearly the community is 'fine-grained'. When Friedman's Test for Randomized Blocks is applied to the rank orders of cover in the twenty stands, the constancy of rank is found to be highly significant, both for the graminoids ($P < 0·001$) and the forbs ($P < 0·001$). Inspection of the original data shows that there is no systematic drift in relative abundances across the plot, analogous to the drift with time in Rabotnov's meadow grassland. The degree of variability in rank within the twenty stands is illustrated in Fig. 2. The ranking is very similar if it is done on mean cover instead of mean rank.

In 1981 the turf at the Sussex site grew about twice as tall as in previous years, almost certainly as a result of reduced grazing pressure in the winter. However, the rank orders of the nine commonest graminoids and eighteen commonest forbs proved to be extremely similar to those found in 1980. According to Kendall's Rank Correlation Test the probability of the correlations arising by chance was <0·01 for the graminoids and <0·001 for the forbs. We thus have a striking consistency of relative abundances at the Sussex site in both space and time.

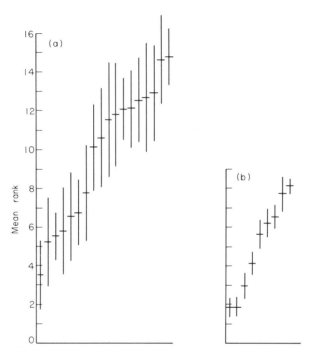

Fig. 2. The mean rank (based on cover) and 95% confidence limits for each of the longer-lived species found in at least 16 out of 20 stands of 0·48 m² in a plot of calcareous grassland at Castle Hill in Sussex (National Grid reference TQ 369069) in late August 1980; (a) forbs, (b) graminoids. Species, reading from left to right, are: (a) *Sanguisorba minor*, *Thymus praecox*, *Cirsium acaule*, *Hippocrepis comosa*, *Asperula cynanchica*, *Leontodon hispidus*, *Lotus corniculatus*, *Plantago lanceolata*, *Hieracium pilosella*, *Succisa pratensis*, *Filipendula vulgaris*, *Scabiosa columbaria*, *Pimpinella saxifraga*, *Plantago media*, *Centaurea nigra*, *Phyteuma orbiculare*, *Ranunculus bulbosus*, *Polygala vulgaris*; (b) *Bromus erectus*, *Brachypodium pinnatum*, *Festuca ovina* + *F. rubra*, *Briza media*, *Carex flacca*, *Koeleria macrantha*, *Avenula pratensis*, *Danthonia decumbens*, *Carex caryophyllea*.

Of the thirty-five species of longer-lived plants intercepted in the Sussex plot, thirty were also intercepted in the Wiltshire plot, plus seven others. The rank order was very similar for the species common to the two sites, both for the graminoids ($P < 0·003$) and the forbs ($P < 0·02$). At the Wiltshire site there was even less difference in the rank orders between 1980 and 1981 than at the Sussex site, for both the graminoids and forbs.

Turning now to a wider comparison, we can relate our rankings obtained in 1980 to those that may be obtained from three sets of forty 1-m² relevées recorded during July–August in chalk grassland in Wiltshire in 1965–68 (Wells 1975). The three sets are

TABLE 4. The numbers of flowering individuals of eight species of short-lived plant in four successive years on two fixed transects (each 50 × 0·5 m) in calcareous grassland at a site in Sussex; the plants are arranged in five groups according to the maximum densities attained*

	South-facing transect				East-facing transect			
	1978	1979	1980	1981	1978	1979	1980	1981
>10 000			Linum 15 000					Rhinanthus 2570
1001–10 000	Linum 4300		Rhinanthus 1300	Rhinanthus 2120	Linum 6100 Rhinanthus 1640	Rhinanthus 3370 Linum 1300	Linum 7700 Rhinanthus 1610	
501–1000	Blackstonia 518	Linum 850 Rhinanthus 723 Centaurium 564	Gentianella 594			Centaurium 625	Gentianella 978	
101–500	Rhinanthus 367 Centaurium 160 Gentianella 140		Blackstonia 263 Centaurium 232	Linum 102	Centaurium 266 Gentianella 208		Centaurium 250	Gentianella 159 Picris 130
11–100	Medicago 32 Picris 13	Medicago 37 Picris 22 Blackstonia 17	Picris 59 Medicago 26	Picris 67 Medicago 28 Blackstonia 26 Centaurium 14	Blackstonia 76 Picris 32 Medicago 22	Blackstonia 29 Gentianella 27 Medicago 25 Picris 24 Carlina 11	Picris 94 Blackstonia 85 Medicago 47	Linum 78 Centaurium 62 Blackstonia 39 Medicago 29
1–10	Carlina 2	Gentianella 5 Carlina 4	Carlina 1	Gentianella 7 Carlina 3	Carlina 8		Carlina 4	Carlina 8

* Linum = L. catharticum, Rhinanthus = R. minor, Blackstonia = B. perfoliata, Centaurium = C. erythraea, Gentianella = G. amarella, Medicago = M. lupulina, Picris = P. hieracioides and Carlina = C. vulgaris.

for grassland in which the most abundant graminoids were respectively *Carex humilis*, *Festuca ovina/rubra* and *Bromus erectus*. We have transformed Wells' cover-abundance values on the Domin scale according to the system of Bannister (1966) in order to bring them to an approximately linear scale, and taken means of the transformed values. The rank orders for the forbs at both our sites are highly correlated with the rank orders found in all three of Wells' grassland types ($P < 0.01$ or 0.001). The ranking of the graminoids at the Sussex site was strongly correlated with that in Wells' *Bromus erectus* type ($P < 0.001$), and that at the Wiltshire site with that in Wells' *Carex humilis* type ($P < 0.01$). There is thus evidence for relatively fixed hierarchies of perennial species in chalk grassland over extensive areas and through fairly long periods of time, wherever management is reasonably constant.

There is much greater variation in the relative abundances of the annuals, biennials and pauciennials. The Sussex site is particularly rich in these plants, possessing at least sixteen species capable of growing in continuous turf. Two of us (P.J.G. & D.K.) have made records of the numbers of flowering plants of eight species in two fixed transects, each 50 × 0.5 m, in four successive years (Table 4). There have been huge fluctuations in the absolute numbers for four of the species. The eight species may be divided arbitrarily on the basis of the maximum densities attained in 25 m^2: *Linum* 7000–15 000; *Rhinanthus* 3000–4000; *Blackstonia*, *Centaurium* and *Gentianella* 500–1000; *Medicago* and *Picris* 50–100; *Carlina* ~10. Our less extensive observations at other sites have documented huge year-to-year variations of the type shown in Table 4, and have suggested that the sequence of maximum densities set out above will probably prove to have general validity.

Communities composed of species of various life-lengths and subject to periodic disturbance

Many natural grasslands suffer periodic disturbance by fire, e.g. the prairies and steppes of the temperate zones, and the savannas of the tropical and subtropical zones. In most cases the relative abundances of species are changed by a fire, and certain species are found only after a fire, while others decrease substantially in abundance in the inter-fire period (Vogl 1974). There is thus a succession of changes between each fire. Comparable 'internal successions' (Curtis 1959) occur where the disturbance is more localized and caused by the activities of animals such as moles, gophers or badgers. Observations which illustrate internal post-fire succession were made for the herb layer in a pine savanna in Georgia, U.S.A., by Lemon (1949). His data for cover are summarized in Table 5. There is a marked hierarchy of abundance in the species that resprout immediately after fire and persist or even increase between fires; all are grasses. *Aristida stricta* and *Sporobolus curtissi* were said by Lemon to 'dominate the herbaceous layer of the pine subclimax in much of southern Georgia and Florida'. There is also a marked hierarchy among the 'fire-followers', with *Andropogon* spp. being particularly prominent, as they are on many disturbed savanna sites in the south-eastern U.S.A. (J. M. Wolf, personal communication). The hierarchy among the fire followers appears to change with time as some species decline sooner after the disturbance than others; clearly more extensive data are needed to establish this point.

We thus have to explain for periodically disturbed communities the fact that some species are generally more abundant than others within both the early successional and late successional assemblages.

TABLE 5. The percentage ground cover of two groups of species in the herb layer at different sites in a pine savanna in Georgia (from Lemon 1949); each site was sampled by eighteen or more plots of 0·001 acre (4·05 m^2)

	Number of growing seasons since last fire		
	1	2	8
Species persisting or increasing after fire			
Sporobolus curtissi	10·0	10·2	9·4
Aristida stricta	8·8	10·0	6·2
Ctenium aromaticum	1·8	1·1	1·9
Sporobolus floridanus	0·4	0·4	1·0
Muhlenbergia expansa	0	0·1	0·4
Fire followers			
Andropogon stolonifer	2·9	2·4	0·9
A. glomeratus + virginicus	2·7	4·2	1·6
Sorghastrum secundum	2·4	2·2	0·3
Sporobolus teretifolius	1·9	0·1	1·0
Aristida virgata	1·2	1·4	0·2
At least 16 species (6 grasses, 4 other monocots, 5 dicots, 1 *Lycopodium*)	6·7	5·7	4·1
At least 10 species (3 grasses, 2 other monocots, 5 dicots)	2·6	0·4	0

A TENTATIVE BASIS FOR UNDERSTANDING

Communities of annuals

The control of relative abundance in this kind of community has been considered critically by Torssell, Rose & Cunningham (1975), on the basis of ideas put forward earlier by Harper and de Wit. The model of Torssell *et al.* involves three 'filters' corresponding to successive stages of the life-cycle: 'germination and establishment' (in fact from seed production to seedling establishment), 'competition' (seedling growth to maturity) and 'seed production'. We prefer to name the second filter 'interference', because it is important to appreciate that plants compete for a place in the landscape at all stages of the life-cycle (cf. Salisbury 1929). It is often useful to subdivide the developmental phases, and produce more filters, as in Fig. 3. It is most unlikely that with any two species any filter will be neutral. This point is well shown by the pair of species studied in detail by Torssell *et al.* (Fig. 3). The legume *Stylosanthes* has a higher rate of seed survival in the face of predation by ants (Mott & McKeon 1977), a higher rate of germination and a greater drought resistance as a young seedling; the grass *Digitaria* has a higher relative growth rate and is favoured in the interference phase, and later on it produces more seeds per unit weight of plant.

A similar analysis was made earlier by Marshall & Jain (1967) for mixtures of *Avena barbata* and *A. fatua* on grazing lands in California. The proportion of *A. fatua* to *A. barbata* was found to increase from the seed to the seedling stage, and again from the seedling to the adult stage, but to fall from the adult to the seed stage. These changes can be explained by the following observations. *A. fatua* produces larger seeds than *A. barbata* and fewer are lost to the subsoil; then seeds of *A. fatua* germinate faster and the seedlings generally have a higher success in establishment, but the adult plants produce fewer seeds per unit weight of plant and the seeds ripen more slowly. As with *Stylosanthes* and

Digitaria, changes in relative abundance from place to place and year to year result from the impact of the microhabitat and weather on all stages of the life-cycle.

Unfortunately the generalized filter in the interference phase of the model of Torssell *et al.* obscures certain important complexities. The outcome of interference, at least in certain species-pairs, depends on both the overall density ('density-dependent effects') and on the relative abundances of the species ('frequency-dependent effects'). Both kinds of effect were illustrated in glasshouse experiments with *Avena barbata* and *A. fatua* by Marshall & Jain (1969). It remains for the extent of these effects to be studied under field conditions;

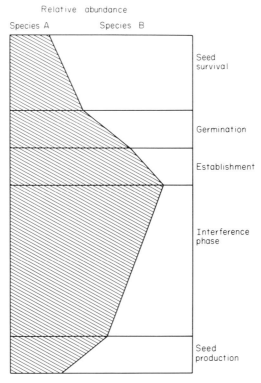

FIG. 3. A graphical representation of changes in relative abundance of two species of annual grassland plant through one annual cycle, based on the scheme of Torssell *et al.* (1975) and showing the generalized relationship between *Stylosanthes humilis* (as species A) and *Digitaria ciliaris* (as species B).

filters other than that for interference may also show frequency-dependence, e.g. seed survival in the face of predation by ants.

Torssell and colleagues were chiefly concerned to explain the balance between the commonest native grasses and the introduced legume, but their mode of analysis could equally well be used to explain the greater abundance of *Digitaria ciliaris* compared with such 'minor grasses' as *Pennisetum polystachyon*. Unfortunately information on the regeneration characteristics of the minor grasses is lacking. However, in a part of California with assemblages of annuals like those studied by Talbot *et al.* (our Table 1), a study of the populations of common and rare species was made by Heady (1958) over a period of 5 yr. He found that the species most abundant at maturity (*Bromus hordaceus* (as *B. mollis*), *B. rigidus* and *Erodium botrys*) had a much higher level of survival between establishment (December) and maturity (June) than the species less abundant at maturity: 75% *v.* 32% averaged over very numerous 6·5-cm^2 samples at two sites of 149 m^2. The less abundant species thus must retain their place by performing relatively better at some other stage of the life-cycle. The extent to which seed may remain dormant in the soil from season to season must also complicate any analysis of this type. Once we take into account spatial distribution, and the patchiness which is usual in these communities (cf. Fig. 1), differences between species in dispersability also become important.

An interesting variant of the kind of community considered so far occurs where the adult plants are very sparse, and there is little obvious interference between plants. In such cases appearances can be misleading, as shown by the following data for the four species we have studied on a sand dune system (Table 2). In early January 1980 the numbers of seedlings were recorded in all 1000 quadrats of 100 cm^2. The records for the *Cerastium* and *Erophila* have been lumped because of uncertainties in identification. The numbers of adults (all flowering) were recorded in the same quadrats in June. It is unlikely that many new plants emerged after the census in January. The mean survivals were as follows: *Valerianella* 41%, *Myosotis* 27%, *Cerastium* + *Erophila* in area Y (where the ratio at maturity was 5 *Cerastium* : 1 *Erophila*) 26%, and *Erophila* + *Cerastium* in area X (2·5 *Erophila* : 1 *Cerastium* at maturity) 19%. The decline in survival parallels a decline in plant size and seed weight. Mean seed weights (μg) are as follows: *Valerianella* 800, *Myosotis* 170, *Cerastium* 80 and *Erophila* 30. Survival was negatively density-dependent for *Valerianella* and for *Cerastium* + *Erophila*, but not for *Myosotis*. In fact, density-dependent effects in *Myosotis* are seen in fecundity rather than in survival, as found for another winter annual (*Vulpia fasciculata*) by Watkinson & Harper (1978). Survival of *Valerianella*, the species with the largest seedlings, was not affected by the density of seedlings of other species, but that of *Myosotis* was, and so was that of *Erophila* + *Cerastium* in area X. There was no effect of density of other species on the survival of *Cerastium* + *Erophila* in area Y; we see no obvious explanation of this result. Clearly the density-dependent effects within species, and the effects of other species on the survival of two of the three smaller-seeded species, demonstrate the reality of interference in even a very sparse community.

The fact that *Erophila* is generally more abundant than *Valerianella* in areas with little moss cover (Table 2) may be explained as follows. *Erophila* is the only one of the four species to produce larger plants and more seeds per plant in sandy areas than in mossy areas. It appears to establish better on bare sand than on moss, perhaps because its tiny seedlings are more readily desiccated in tufts of moss. *Erophila* thus finds more suitable micro-sites for establishment in areas of low moss cover. *Valerianella* persists in the community because its seedlings, which become established on moss tufts, are relatively large, and are able to suppress the growth of any *Erophila*, without being affected themselves (at least in terms of survival) by the *Erophila*. The *Cerastium* and *Myosotis* may be tentatively considered as intermediate in these respects between *Erophila* and *Valerianella*.

It is probably significant that in an environment which is generally unfavourable for plant growth (the sand dune system) the most abundant species is the one with lowest survival as an established plant and least potential for interference, but greatest resistance to the environmental hardships during establishment, while in an environment more favourable for plant growth (the Californian pasture) the most abundant species are those with greatest survival as established plants, and perhaps greatest potential for interference.

The explanation of the relative abundances of *Erophila* and *Valerianella* just given assumes a constant low level of moss cover. In practice areas of $1-10$ m^2 are continually becoming more and less moss-rich as a result of local variations in the rate of accretion of sand. *Erophila* would probably become very rare in the system, were it not for the periodic 'disturbance' of sand accretion. Witness the decline in its relative abundance in area Y from 1979 to 1980, and its resurgence in 1981 when much new sand was deposited.

In summary, if a species is persistently found at a relatively low level of abundance, it must be that either the micro-sites suitable for its establishment are less common than those that are suitable for more abundant species, or that it is less capable of occupying all the micro-sites suitable for it, e.g. through low seed production, ineffective dispersal, heavy seed predation or poor seedling survival. The problem is to explain the indefinite persistence of the less abundant species. If there are relatively few micro-sites suitable for it, it may still persist if it can invade more often (more seed per unit of adult plant weight, seed more dispersible in space or time) or have a greater potential for interference in the short term. If the less abundant species is less capable of reaching all the appropriate micro-sites, it may still persist if it has a greater potential for interference at such sites. We shall be able to substantiate these points as generalizations only when the 'regeneration niches' (Grubb 1977) of many plants have been thoroughly explored.

So far we have assumed that all the species in a community are indeed capable of indefinite persistence in it, but in practice this may well not be so. In phytosociological work it is usual to find transgressors, i.e. species which are not typical of the community but which have somehow invaded in a few places, and are unlikely to be able to reproduce themselves. The small populations of such species in the 'wrong' communities have to be 'topped up' from time to time by influx from neighbouring sites, where the species do persist indefinitely.

Communities composed of plants of various life-lengths and subject to continual disturbance

In principle the kind of analysis used by Torssell and others for annuals may be applied here also, except that differences between species in longevity and potential for vegetative spread become important. Other things being equal, longer-lived species and those whose adults have a greater mean size, must be more abundant. Six characteristics of species may then be isolated for preliminary study: dispersability, frequency of establishment from seed, potential to resist interference as seedlings facing adults, potential to bring about interference as adults facing adults, longevity and mean area occupied by an adult. To be at the top of a hierarchy such as we have found in chalk grassland may require a high rating for all these characteristics other than dispersability, while the possession of a low rating for any of the characteristics other than dispersability may be enough to bring the species near the bottom of the hierarchy. As yet we have little critical information for the species shown in Fig. 2. However, the most abundant species are certainly all long-lived as judged by records of mapped plants over 8 yr; they also spread considerably from one seedling and are often seen as newly established seedlings. Some at least are effective as seedlings in resisting interference from artificial swards of *Festuca rubra* on chalk grassland soil: *Bromus erectus*, *Sanguisorba minor* and *Leontodon hispidus* (P. E. Gay, personal communication). In contrast relatively short life and limited vegetative spread probably contribute materially to the low ranking of *Pimpinella saxifraga* (unpublished observations) and *Ranunculus bulbosus* (cf. Sarukhán & Harper 1973). A low potential as a seedling to resist interference from *Festuca rubra* may be important in determining the low ranking of *Phyteuma orbiculare* (cf. Fenner 1975). Clearly much more observational and experimental work is needed to explain the hierarchies exemplified in Fig. 2. As yet we have little idea of the circumstances under which the less abundant species are advantaged, and so enabled to keep their place in the community. In some cases frequency-dependent effects may contribute to the determination of relative abundance of pairs of species, as shown for *Anthoxanthum odoratum* and *Phleum pratense* by van den Bergh & de Wit (1960), and for *Anthoxanthum odoratum* and *Plantago lanceolata* by Berendse (1981).

Where the hierarchies are much less fixed, as in Rabotnov's river meadow, the same characteristics are important but the competitive balance is more sensitive to weather or slight changes in management, and marked changes in rank almost certainly persist for years after the key environmental events have passed away. Of the marked transitions shown in Table 3, the marked rise in *Bromus inermis* and *Elymus repens* in 1958–59 can be related to the hot dry summer of 1959 (cf. Rabotnov 1974). The reciprocal changes in rank of several legumes and two grasses in 1955–56 and 1957–58 may be tentatively related to good and bad seed years for the legumes (Rabotnov 1966). In other cases the grass-legume balances can be controlled very closely by appropriate cutting or grazing regimes and fertilizer treatments (Haynes 1980). However, we have to accept that at the present time the effects of weather on relative abundance (apart from extreme drought) are generally obscure, probably in part because of complex lag effects from previous years. Van den Bergh (1979) has provided a summary of what has been achieved in interpreting year-to-year changes in grassland.

We shall publish elsewhere detailed studies on the populations of annuals, biennials and pauciennials in chalk grassland. Our evidence from following the fates of thousands of seedlings suggests that although they suffer marked interference from the matrix-forming perennials, they interfere with each other (both within species and between species) to only a limited extent. Here we are concerned only to try to explain the differences between species in the maximum densities attained. At first sight the major explanation seems to be the size of the mature individual. The mature plants of *Linum catharticum*, the species which reaches the highest densities, often spread their branches over only 1–3 cm^2 and interpenetrate other plants of the same species. In contrast, the flowering rosettes of *Carlina vulgaris*, which attain very low densities, occupy 15–25 cm^2 and do not interpenetrate their own kind. The *Rhinanthus* and Gentianaceae are intermediate in this respect, while *Medicago* and *Picris* resemble *Carlina* in area occupied per flowering individual. The number of gaps in the turf big enough to allow maturation of a flowering plant may thus be expected to be successively less for *Linum* > *Rhinanthus* and Gentianaceae > *Medicago*, *Picris*, *Carlina*. It may also be significant that the three species with the lowest maximum densities take 2–4 yr to mature rather than the 1–2 yr for the other species, and they presumably cannot increase their population size as quickly during a run of favourable years. However, other factors to do with the survival of seeds and the incidence of conditions inducing germination also seem to be important. In the case of *Carlina* a major limitation on population size is seed predation by small mammals (Greig-Smith & Sagar 1981).

Communities composed of plants of various life-lengths and subject to periodic disturbance

In any community which displays a succession of most abundant species between disturbances, it is plain that the relative abundances of early, mid and late successional species at any one place and time will depend on the extent of the last disturbance and the length of time since it occurred. Certain authors have tried to make analyses of such successions in terms of first-order Markov chains, but van Hulst (1979) has pointed out that successions generally cannot be analysed usefully in such terms; in order to forecast the future vegetation at a site it is necessary to know not only its present state but also its past history. A more useful approach is a mechanistic one, such as the analysis of 'vital attributes' put forward by Noble & Slatyer (1980). They emphasized the importance of the mode of arrival and persistence at a site, the ability to invade at different stages in succession, and three key time functions: time to first fruiting, length of life of adults, and length of life of dormant seed in the soil. They showed that the patterns of relative abundance among the trees in certain parts of the world subject to periodic disturbance by fire could be substantially explained in terms of the 'vital attributes' of the various species. It appears that no such analysis has been attempted for an herbaceous community, such as the herb layer of a pine savanna for which data are given in Table 5. Any adequate analysis must explain not only the proportions of early and late successional

TABLE 6. The occurrences and mean biomasses (g m^{-2}) of grasses and forbs in nine random 1-m^2 samples in roadside *Arrhenatheretum* in south-eastern Cambridgeshire; recorded in July. Material was cut 8 cm from the ground. (For further details see Grubb 1982)

Grasses			Forbs		
9	*Arrhenatherum elatius*	301	7	*Anthriscus sylvestris*	19
9	*Elymus repens*	33	7	*Galium aparine*	2·3
9	*Festuca rubra**	15	6	*Convolvulus arvensis*†	7·2
8	*Poa pratensis*	13	3	*Plantago lanceolata*	4·4
7	*Dactylis glomerata*	16	3	*Heracleum sphondylium*	4·2
4	*Bromus sterilis*	3·9	3	*Achillea millefolium*	0·7
4	*Phleum pratense*	3·0	3	*Glechoma hederacea*	0·3
4	*Trisetum flavescens*	1·6			
4	*Lolium perenne*	0·6			
4	*Agrostis stolonifera*	0·3			

* Not flowering in 4/9.
† Not flowering in 6/6.

species at a site but also the hierarchy of relative abundance within each of these groups (cf. Table 5).

It now seems likely that periodic localized disturbance, and the reactions of various species to it, contribute significantly to the determination of relative abundances in communities where it is not obviously so at first sight, including some that are also subject to continual disturbance such as mowing. As an example we take roadside *Arrhenatheretum* in Cambridgeshire; it receives variable management, but is rarely cut more than once a year. The community occupies roadside strips 0·5–2·0 m wide, and is often found between a more frequently mown shorter grassland at the roadside edge and a hedge. There are marked seasonal changes in relative abundance; the frequencies and biomasses of the species occurring in at least three out of nine samples chosen at random in late July are shown in Table 6.

An experiment was set up in 1973 at the Cambridge Botanic Garden to try to explain the relative abundances of four species in the community in question: *Arrhenatherum elatius*, *Dactylis glomerata*, *Festuca rubra* and *Plantago lanceolata*. Each species was grown by itself, in mixture with *Dactylis* and in mixture with all three other species. Seedlings were set out in replicate plots of 1·5 × 1·5 m and a guard row of *Dactylis* was set out around the collection of plots. During 8 yr there has been little invasion by other species. After a year the plots were harvested in July at a height of 8 cm, and the results offered little by way of explaining the relative abundances at field sites. In the four-species plots the yields were *Dactylis* > *Plantago* > *Arrhenatherum* > *Festuca*. Moreover, *Dactylis* out-yielded *Arrhenatherum* not only in the pure plots but also when they were grown together. In the following year a decreased yield was recorded in only one out of six plots for *Arrhenatherum*, but in almost all plots for other species. During the great drought of 1975–6 there was little growth and the plots were not harvested in 1976 or 1977.

By 1978 some remarkable changes were found. Over about 60% of the length of the guard rows *Dactylis* had been totally replaced by *Arrhenatherum*, and the *Arrhenatherum* had invaded all the plots not sown to that species. However, the two plots sown to *Arrhenatherum* had not been invaded by any of the other species. By 1979, when a harvest was taken, *Arrhenatherum* was the most abundant species in all plots within *c*. 2 m of plots sown originally with that species. In the four-species plots the yields were *Arrhenatherum* ≫ *Festuca* > *Dactylis* > *Plantago*; in 1979–1981 the *Festuca* increased in these plots, while the *Dactylis* showed little change. As a result of this experiment it appears that the *Dactylis* and *Plantago* (at least) owe their place in the community to some form of periodic disturbance, the exact nature of which is uncertain.

At the first harvest the greater potential for interference in plants less abundant in the field, i.e. *Dactylis* and *Plantago*, seemed surprising, but if a plant is disadvantaged at some stage between disturbances, and it is to remain in the community, it must be advantaged at some other stage, rather as we saw that annual plants able to occupy fewer micro-sites must be able to invade them more effectively or have a greater potential for interference. It is particularly interesting that the less abundant grasses in a prairie in Missouri have been found in glasshouse experiments to have a greater potential for interference than the more abundant species (Rabinowitz 1981). The less abundant species also have more dispersible seed (Rabinowitz & Rapp 1981), and the adult plants are probably shorter-lived (D. Rabinowitz, personal communication).

Experiments in which single species are removed from a community that has not recently been disturbed are likely to give an opposite impression of what might loosely be called the potential for interference in more and less abundant species. Thus Silander (1976) removed single species or pairs of species from five community types on the coast of North Carolina (low marsh, high marsh, primary dune, rear dune and inter-dune (swale) grassland), and recorded the reactions of the remaining species over a year. The pattern of semi-natural disturbance experienced by the communities concerned probably includes periodic burning and opening up by animals. Most of the experimental effects involving the single most abundant species were one-sided, suggesting that the more abundant species were occupying as much room as they could. Thus in swale grassland removal of the most abundant species *Muhlenbergia capillaris* (which had a cover of 48% in the control) led to significant increases in the cover of all the other six species having a cover of 0·2% or more in the control, but removal of less abundant species (or pairs of species) with 11, 10 and 3% cover in the control led to no signficant change in the cover of *Muhlenbergia*. In contrast, the effects lower down the hierarchy tended to be reciprocal. Removal of *Andropogon scoparius* + *Eragrostis pilosa* (together 11% in the control) led to a significant increase in cover of *Hydrocotyle bonariensis* + *Solidago sempervirens* (10% in control), and vice-versa. However, we should not read too much into these results. A statistically significant change of 3–6% cover is easier to obtain when the control cover is 3–6% than when it is 48%. Moreover, species which spread vegetatively usually expand more quickly than those which spread by seed, and the result of perturbation may be appreciably different after 2 yr from what it is after only one. Nevertheless it does appear likely that there is a hierarchy of potential for occupancy in the late post-disturbance phase. This hierarchy appears to depend on such features as potential for forming tall, dense and widespread individuals. It should not be confused with the hierarchy that we should expect on newly cleared ground, when the generally less abundant species ought (according to the results for the English *Arrhenatheretum* and American prairie) to show greater potential for interference, e.g. by invading more quickly, showing greater tolerance of desiccation in an open site, and having a higher relative growth rate. With the passage of time after a disturbance the potential for interference on the part of the late successional species becomes realized.

What is needed now for a number of herbaceous communities is a combination of both the synthetic type of experiment done for the *Arrhenatheretum*, and the selective removal experiment done for the coastal communities in North Carolina.

CONCLUDING REMARKS

So little advance has been made since Watt (1961) drew attention to the potential interest of trying to understand the control of relative abundance in natural and semi-natural vegetation that much of what we have written in this article has had to be preliminary or speculative in nature. Because of the limitation on space we have discussed only ranking of species, and written almost nothing about the pattern of distribution of abundance in plant communities, about which much has been written by others (cf. Whittaker 1977). Despite a wealth of descriptive studies in that area, little has been achieved in terms of a mechanistic understanding. By confining ourselves to herbaceous communities, we have failed to draw attention to the advances in understanding made by some who have modelled forest succession after disturbance, e.g. Botkin, Janak & Wallis (1972) and others since. Our course has been deliberate. We wish to point out how limited is our understanding even of simple communities, and the challenge that is there for those who are prepared to make painstaking quantitative descriptive studies of the life-cycles of coexisting species, together with appropriate experiments.

Finally, in the context of nature conservation, we may agree with Watt (1971, p. 137) that the 'identification of factors controlling floristic composition' is 'one of the most fascinating as well as urgent problems confronting the plant ecologist'.

ACKNOWLEDGMENTS

It gives the senior author great pleasure to acknowledge his debt to Dr Watt for inspiring his interest in the control of relative abundance. We all thank the Nature Conservancy Council for permission to work on their reserves, and Drs J. Antonovics and D. Rabinowitz for access to unpublished results. Dr E. I. Newman's critical comments on an early draft have been of great value.

REFERENCES

Bannister, P. (1966). The use of subjective estimates of cover abundance as the basis for ordination. *Journal of Ecology*, 54, 665–674.

Berendse, F. (1981). *Competition and equilibrium in grassland communities.* Proefschrift, University of Utrecht.

Bergh, J. P. van der (1979). Changes in the composition of mixed populations of grassland species. *The Study of Vegetation* (Ed. by M. J. A. Werger), pp. 59–80. Junk, The Hague.

Bergh, J. P. van der & Wit, C. T. de (1960). Concurriente tussen Timothee en Reukgras. *Mededelingen Instituut voor biologisch en scheikundig onderzoek van landbouwgewassen,* **121,** 155–165.

Botkin, D. B., Janak, J. F. & Wallis, J. R. (1972). Some ecological consequences of a computer model of forest growth. *Journal of Ecology,* **60,** 849–872.

Curtis, J. T. (1959). *The Vegetation of Wisconsin.* University of Wisconsin Press, Madison.

Fenner, M. (1975). *Factors affecting the distribution of strict calcicoles.* Ph.D. dissertation, University of Cambridge.

Greig-Smith, J. & Sagar, G. R. (1981). Biological causes of local rarity of *Carlina vulgaris. The Biology of Rare Plant Conservation* (Ed. by H. Synge), pp. 389–400. Wiley, Chichester.

Grime, J. P. (1979). *Plant Strategies and Vegetation Processes.* Wiley, Chichester.

Grubb, P. J. (1977). The maintenance of species-richness in plant communities: the importance of the renegeration niche. *Biological Reviews,* **52,** 107–145.

Grubb, P. J. (1982). An experimental approach to understanding the control of relative abundance in roadside Arrhenatheretum. *Journal of Ecology,* **70,** in press.

Harper, J. L. (1961). Approaches to the study of plant competition. *Mechanisms in Biological Competition* (Ed. by F. L. Milthorpe), pp 1–39. *Symposia of the Society for Experimental Biology,* 15.

Harper, J. L. (1977). *Population Biology of Plants.* Academic Press, London.

Haynes, R. J. (1980). Competitive aspects of the grass-legume association. *Advances in Agronomy,* **33,** 227–261.

Heady, H. F. (1958). Vegetational changes in the California annual type. *Ecology,* **39,** 402–416.

Hulst, R. van (1979). On the dynanics of succession: Markov chains as models of succession. *Vegetatio,* **40,** 3–14.

Lemon, H. C. (1949). Successional responses of herbs in the longleaf-slash pine forest after fire. *Ecology,* **30,** 135–145.

Marshall, D. R. & Jain, S. K. (1967). Cohabitation and relative abundance of two species of wild oats. *Ecology,* **48,** 656–659.

Marshall, D. R. & Jain, S. K. (1969). Interference in pure and mixed populations of *Avena fatua* and *A. barbata. Journal of Ecology,* **57,** 251–270.

McCown, R. L. & Williams, W. A. (1968). Competition for nutrients and light between the annual grassland species *Bromus mollis* and *Erodium botrys. Ecology,* **49,** 981–990.

Mott, J. J. & McKeon, G. M. (1977). A note on the selection of seed types by harvester ants in northern Australia. *Australian Journal of Ecology.* **2,** 231–235.

Noble, I. R. & Slatyer, R. O. (1980). The use of vital attributes to predict successional changes in plant communities subject to recurrent disturbances. *Vegetatio,* **43,** 5–21.

Pemadasa, M. A. & Lovell, P. H. (1976). Effects of the timing of the life-cycle on the vegetative growth of some dune annuals. *Journal of Ecology,* **64,** 213–222.

Rabinowitz, D. (1981). Seven forms of rarity. *The Biology of Rare Plant Conservation* (Ed. by H. Synge), pp. 205–217. Wiley, Chichester.

Rabinowitz, D. & Rapp, J. K. (1981). Dispersal abilities of seven sparse and common grasses from a Missouri prairie. *American Journal of Botany,* **68,** 616–624.

Rabotnov, T. A. (1966). Peculiarities of the structure of polydominant meadow communities. *Vegetatio,* **13,** 109–116.

Rabotnov, T. A. (1974). Differences between fluctuations and seasons. *Handbook of Vegetation Science. Part 8. Vegetation Dynamics* (Ed. by R. Knapp), pp. 19–24. Junk, The Hague.

Salisbury, E. J. (1929). The biological equipment of species in relation to competition. *Journal of Ecology,* **17,** 197–222.

Sarukhán, J. & Harper, J. L. (1973). Studies in plant demography: *Ranunculus repens* L., *R. bulbosus* L. and *R. acris* L. I. Population flux and survivorship. *Journal of Ecology,* **61,** 675–716.

Silander, J. A. (1976). *The genetic basis of the ecological amplitude of* Spartina patens *on the Outer Banks of North Carolina.* Ph.D. dissertation, Duke University, Durham, North Carolina.

Sokal, R. R. & Rohlf, F. J. (1969). *Biometry.* Freeman, San Franciso.

Symonides, E. (1979). The structure and population dynamics of psammophytes on inland dunes. III. Populations of compact psammophyte communities. *Ekologia polska,* **27,** 235–257.

Talbot, M. W., Biswell, H. H. & Hormay, A. L. (1939). Fluctuations in the annual vegetation of California. *Ecology,* **20,** 394–402.

Torsell, B. W. R. (1973). Patterns and processes in the Townsville stylo annual grass pasture ecosystem. *Journal of Applied Ecology,* **10,** 463–478.

Torsell, B. W. R., Rose, C. W. & Cunningham, B. R. (1975). Population dynamics of an annual pasture in a dry monsoonal climate. *Proceedings of the Ecological Society of Australia,* **9,** 157–162.

Tutin, T. G., Heywood, V. H., Burges, N. A., Valentine, D. H., Walters, S. M. & Webb, D. A. (1964–1980). *Flora Europaea*, Vols. 1–5. Cambridge University Press.

Vogl, R. J. (1974). Effects of fire on grasslands. *Fire and Ecosystems* (Ed. by T. T. Kozlowski & C. E. Ahlgren), pp. 139–194. Academic Press, New York.

Watkinson, A. R. & Harper, J. L. (1978). The demography of a sand dune annual: *Vulpia fasciculata*. I. The natural regulation of populations. *Journal of Ecology*, **66**, 15–33.

Watt, A. S. (1961). Ecology. *Contemporary Botanical Thought* (Ed. by A. M. McLeod & L. S. Cobley), pp. 115–131. Oliver & Boyd, Edinburgh.

Watt, A. S. (1971). Factors controlling the floristic composition of some plant communities in Breckland. *The Scientific Management of Animal and Plant Communities for Conservation* (Ed. by E. Duffey & A. S. Watt), pp. 137–152. *Symposia of the British Ecological Society*, **11**.

Wells, T. C. E. (1975). The floristic composition of chalk grassland in Wiltshire. *Supplement to the Flora of Wiltshire* (Ed. by L. F. Stearn), pp. 99–125. Wiltshire Archaeological & Natural History Society, Devizes.

Whittaker, R. H. (1977). Evolution of species diversity in land plant communities. *Evolutionary Biology*, **10**, 1–67.

THE EFFECT OF FUNGAL PATHOGENS ON PLANT COMMUNITIES

J. J. BURDON

*Division of Plant Industry, C.S.I.R.O., P.O. Box 1600,
Canberra A.C.T. 2601, Australia*

SUMMARY

Various aspects of the potential impact of pathogens on the genetic structure of individual plant species and on the organization and diversity of whole plant communities are considered.

At an intra-specific level it is shown how coevolutionary interactions between host plants and their pathogens may lead to the establishment and maintenance of genetically diverse plant populations. In such populations individual plants may be protected by qualitative resistance, quantitative resistance, tolerance, or may avoid disease altogether by escaping in time. The population as a whole benefits not only from the resistance possessed by its individual members but also gains a further 'population resistance' benefit from the combined effects of the available protective mechanisms.

At an inter-specific level the effect of disease on competitive interactions between different species is stressed and examples are presented to illustrate how this may lead to increases in the species diversity of such communities.

Finally, the impact of pathogens on the macro- and micro-geographic distribution of host species is considered.

INTRODUCTION

In many plant communities disease incidence is sporadic and for most of the time the only obvious sign of pathogen activity is the occasional blotch or pustule on stem or leaf. Possibly because of this, pathogens have rarely been studied by ecologists as part of natural ecosystems and, consequently, their relevance to the population dynamics and genetics of individual host species and to community diversity in general is poorly recognized (Harper 1977). The main aim of this paper is to consider the ways in which pathogens may influence both the genetic structure of individual populations and the balance achieved between competing sympatric species. In addition, their possible effects on the pattern of distribution of host species within and between different habitats and on the phenology of individual species is also considered. The theoretical bases for these postulated interactions are examined and, wherever possible, evidence for their existence is provided from natural plant communities. Inevitably, though, the heavy bias of plant pathological studies towards agro-ecosystems makes it necessary to draw frequently on this source for examples of the types of interaction being discussed.

A logical place to start in any consideration of disease in plant communities is the effect of disease on single species stands. Freed from external biological constraints, many plant populations have the potential to increase in size until restricted by competition for

0262–7027/82/0300–0099$02.00 © 1982 British Ecological Society

resources in limited supply. The high density, single species stands which result from such unrestricted growth are, because of the presence of large amounts of susceptible tissue and the potentially low loss of inoculum during dispersal, particularly vulnerable to attack by fungal pathogens (Burdon & Chilvers 1975, 1976a). The enhanced activity of pathogens at high plant densities may affect the size of future generations and may eventually result in the incomplete utilization of resources, thereby creating space (*sensu* de Wit 1960) or 'partly vacated niches' (Harper 1969) available for colonization.

Such a situation is likely to force a species to respond in one of two ways to minimize the impact of the pathogen on the population as a whole. Firstly the same species may, through changes in the genetic composition of the population, reduce the impact of the pathogen and recolonize the vacant space. Alternatively, these partly vacated niches may be colonized by other plant species resistant to attack by this pathogen. In virtually all interactions between plants and pathogens at a population level, the final outcome is likely to incorporate both reductions in density and changes in the genetic structure of host populations. For simplicity these two possibilities are treated separately here, although, as we shall see, interactions at intra- and interspecific levels have the same essential features.

PATHOGENS AND THE GENETIC STRUCTURE OF HOST POPULATIONS

Theoretical considerations

For any individual in a population the occurrence of disease generally results in a lowering of its vigour and competitive ability relative to its neighbours. As a result, when a uniform host population with a single gene for resistance is confronted, for the first time, by a uniform pathogen population possessing the complementary virulence gene (that is the plant population is uniformly susceptible), the chance mutation producing a novel resistance allele that enables a host to escape disease through enhanced resistance will clearly be favoured. Such individuals will possess a greater fitness than their susceptible neighbours, will reproduce more freely and will, consequently, increase in frequency in the population. While this newly resistant genotype remains at a low frequency in the population, the selective pressure on the pathogen population favouring the emergence of a new race capable of attacking this novel resistance will remain low. However, as the frequency of the novel genotype increases further, the selective pressure on the pathogen population will also increase until a chance mutation produces a new race possessing a new virulence gene. When this happens the selective advantage previously enjoyed by the mutant host genotype is lost, its competitive position relative to the original genotype is weakened and its frequency in the population declines. With continued pressure being exerted on the plant population by the diversified pathogen population, the emergence of yet another resistance allele will be favoured. So long as the pathogen remains a significant part of the host plant's environment this coevolutionary interaction will continue, resulting eventually in a host and pathogen population of considerable diversity. Such a situation has been modelled in detail by Person (1966) who proposed that a balance was achieved and maintained between the host and pathogen populations through a series of cyclical polymorphisms between specific oligogenic resistance factors (resistance controlled by a few genes) in the host and complementary virulence characters in the pathogen.

More recently, increasing interest in the use of a range of different resistance genes within the one crop as a strategy for disease control in agricultural systems had led to considerable speculation concerning the interactions between such mixtures and the virulence structure of the pathogen (Browning & Frey 1969; Leonard 1977; Marshall 1977; Barrett 1978; Marshall & Pryor 1978). Such studies tend to agree that in the sort of simplified situation proposed by Person (1966) individual pathogen races are likely to become more and more complex until a race capable of attacking all host genotypes equally will predominate. However, such a frightening scenario may be averted if unnecessary genes for virulence in the pathogen or resistance in the host impose a fitness cost (Leonard 1977; Barrett 1978; Marshall & Pryor 1978) or if other disease control strategies are taken into account. When these restrictions are applied the pathogen races which predominate tend to be ones with intermediate levels of virulence.

Practical evidence: the complexity of coevolved systems

Such models and theoretical speculations are all very well but do such interactions actually occur in nature? Despite a lack of many examples, the answer to this question is generally yes, with evidence coming from a range of sources. Thus the primary and secondary centres of diversity of plant species, where host and pathogen have generally been in contact for long periods of time, have always been particularly fruitful places to search for resistance to a wide range of diseases (Harlan 1976). For example, collections of barley from Ethiopia have shown high resistance to eight different fungal and viral diseases; collections of wheat from Turkey also show considerable disease resistance, with one collection in particular showing resistance ranging from usable tolerance to almost complete immunity to at least fifty-one races of a total of six different fungal pathogens (Harlan 1976).

At the other extreme, populations distant from the centre of origin may, due to a lack of selective pressure, have lost resistance genes or failed to accumulate those occurring naturally through random mutation. Thus North American chestnut populations, evolving in isolation away from their ancestral south-east Asian home, were under no selective pressure to accumulate resistance to pathogens associated with the Asian centre of diversity. Consequently, when one of these pathogens, namely *Endothia parasitica*, was accidentally introduced the existing population was highly susceptible (Burdon & Shattock 1980). North American five-needle pine species have also been found to be highly susceptible to another accidently introduced pathogen, *Cronartium ribicola* (Stern & Roche 1974), although in this case a very low percentage of trees have been found to be resistant.

So far in this discussion no precise description or definition has been given of the type of resistance being considered. This has been deliberate. While much of the resistance used by plant breeders in the production of resistant varieties of crop plants is characterized by the failure of the pathogen to reproduce at all, in natural plant-pathogen interactions a wide range of types of resistance may be found. The recognition of the existence of these various forms of resistance is of considerable importance as it seems likely that in many cases it is the occurrence of this range of different types of resistance in a host population which leads to long-term stability in the appropriate host-pathogen combination (Nelson 1978).

The types of disease resistance encountered in natural plant populations may be classified into two major types—active and passive. The passive resistance mechanisms of interest here are those in which past interactions between host and pathogen have led to the plant population responding through a process of avoidance or disease escape, that is, spatial or temporal diversity. Active disease resistance mechanisms, on the other hand, occur as a result of an ongoing active response of the host to the pathogen and incorporate resistance factors which either reduce the rate of colonization and reproduction of the pathogen following successful infection or prevent it altogether. Such resistance may range, on a continuous scale, from high resistance characterized by the production of small necrotic lesions and no sporulation to extreme susceptibility characterized by the production of many, large prolific pustules after short development periods. While there is little evidence of any fundamental genetic difference between resistance which totally prevents sporulation and resistance which reduces the rate of spore production (both may be under either oligogenic or polygenic control), it is often convenient to classify active resistance into these two types (qualitative and quantitative respectively) as they *do* differ in their effect on the epidemiology of the pathogen and thus in their potential long-term effect on host-pathogen interactions. Resistance which reduces the speed at which reproduction occurs reduces the rate of disease increase, while resistance which prevents sporulation reduces the amount of effective initial inoculum by the proportion of resistant plants in the population. The former type of resistance serves then, to reduce the rate at which disease builds up while the latter postpones an epidemic in time. A third form of active interaction between host and pathogen which appears to play a role in some plant communities is that of disease tolerance. In interactions with tolerant plant genotypes, the pathogen sporulates freely (the plant giving the appearance of a typical susceptible or highly susceptible reaction) but the impact on the host, as measured in terms of its reproductive output, is small.

The continuous range in levels of active resistance found within host plant populations is illustrated in the resistance structure of a single natural population of *Trifolium repens* to two common foliar pathogens, *Cymadothea trifolii* and *Pseudopeziza trifolii* (Burdon 1980). Within a random sample of fifty *T. repens* plants the reaction of individual clones to infection and subsequent disease development was extremely variable. Some clones were highly resistant while others were highly susceptible to the pathogen isolate used. However, the response of the population as a whole to the two pathogens was quite different. The spectrum of resistance to *C. trifolii* was normally distributed, with the bulk of the population showing intermediate levels of resistance. By contrast, resistance to leaf spotting by *P. trifolii* was strongly skewed with the majority of the population showing a high degree of resistance. While the reasons for these differences are not fully understood it is possible that they reflect the relative impact of these two pathogens on plant fitness.

Similar within-population differences have been found in resistance to *Erysiphe graminis* f. sp. *hordei* in Israeli wild barley (*Hordeum spontaneum*) populations, to *Puccinia coronata* and *P. graminis* f. sp. *avenae* in wild oat populations in Australia (*Avena barbata* and *A. fatua*) and Israel (*A. barbata* and *A. sterilis*), and to *Phakopsora pachyrhizi* in four native Australian species of *Glycine* (Wahl *et al.* 1978; Burdon, unpublished data; Dinoor 1970; 1977; Burdon & Marshall 1981, and unpublished data respectively). Moreover, in most of these studies, evidence was gathered to show that, in at least some of the populations, many different qualitative and quantitative resistance genes were deployed. In the Israeli wild oat studies the use of six different races of *Puccinia coronata* has uncovered at least six qualitative resistance genes occurring together in the same population, while in the Australian study the use of five different races of *P. coronata* and four different races of *P. graminis* f. sp. *avenae* has uncovered three and four different qualitative resistance genes respectively within the one host population. Furthermore, varying levels of quantitative resistance have been found to be distributed between individual members of many of these populations.

These qualitative and quantitative resistance factors are distributed apparently haphazardly so that some host plants are strongly defended, by a combination of qualitative and quantitative resistance, against most of the pathogen races likely to occur, other individuals are well protected against some pathogen races but vulnerable to others, while still other hosts are susceptible to virtually all pathogen races.

Differences may also occur between the resistance structure of populations of the one species. Thus different Australian populations of both *Avena barbata* and *A. fatua* have been found to vary in the relative proportions of the population showing qualitative and quantitative resistance (Burdon, unpublished data). Similar results have also been obtained in *A. sterilis* populations where the frequency of quantitative and qualitative resistance factors varies markedly from population to population (Segal *et al.* 1980; Dinoor 1977). In these populations host tolerance of infection has also been implicated as a further twist in the coevolutionary interaction of host and pathogen. However, while tolerant individuals are known to occur in some populations (Wahl 1958; Segel *et al.* 1980), their relative frequency is unknown.

Avoidance of disease in either time or space is another way in which a plant population may respond to pathogen pressure. Avoidance in space may be achieved through simple variations in either micro- or macro-geographic distribution and may occur without genetic change in the host population. This type of avoidance is considered later. Disease avoidance in time, on the other hand, requires a temporal separation of the optimal growing conditions of host and pathogen. The most obvious way this may come about is through early flowering and ripening, thus ensuring that important stages of plant growth are passed before environmental conditions favourable to the increase of the pathogen occur. Providing irrefutable evidence of such a relationship between earliness and later high disease levels is virtually impossible. However, circumstantial evidence may be obtained by comparing the disease control strategies employed by closely related species attacked by the same pathogen.

Results of a recent study of co-occurring populations of *Avena barbata* and *A. fatua* in southern New South Wales are consistent with the hypothesis that earliness may be a disease avoidance strategy (Burdon, unpublished data). An analysis of the resistance structure of five populations has shown that *A. fatua* populations consistently possess more qualitative and quantitative resistance (to five races of *Puccinia coronata*) than do co-occurring *A. barbata* populations ($0.1 > P > 0.05$). These observations were supplemented with a glasshouse experiment to examine the response of these species to disease caused by *P. coronata* (crown rust). Each species was grown in pairs with only one member of each pair being infected, the other being kept relatively disease free by fungicide application. Plants were infected either at the three-leaf stage (early) or as the *A. barbata* plants headed (late). When infection was early, the synergistic interaction of disease and competition from a healthy partner produced considerable reductions in dry weight, number of fertile tillers and average seed size in both species (Table 1). However, when infection was late only *A. barbata* was significantly affected even though it was starting to head when infected whereas the *A. fatua* plants were 2–3 weeks more immature and therefore exposed to attack for longer. The data in Table 1 are the mean of six replicates of the infected members of each pair. Controls were obtained from plants growing in pots in which neither individual was infected.

Crown rust occurs in southern New South Wales throughout the growing season but severe infections are rare before mid-spring. The inability of *A. barbata* to withstand attack even as late in its development as heading and the low levels of active resistance factors found in these populations suggest that the disease control strategy adopted by this species is one of avoidance, flowering as it does 2–3 weeks earlier in the spring than *A. fatua*. *Avena fatua*, on the other hand, seems to rely on a greater level of qualitative and quantitative resistance coupled with its greater ability to withstand the effects of disease. Workers studying the interaction between *A. sterilis* and *Puccinia coronata* in Israel have suggested a similar utilization of disease avoidance by some populations of this species (Segal *et al.* 1980).

Overall, four different components of defence (viz., qualitative resistance, quantitative resistance, tolerance and disease avoidance or escape) may be recognized as occurring in, and contributing to, the stability of natural plant populations. Moreover, the contribution that some of these mechanisms make to disease control in populations surpasses their protective effect on the individual alone. Thus the presence of qualitative resistance genes protects not only those individuals possessing them against pathogens with inappropriate

TABLE 1. The impact of moderate levels of disease caused by *Puccinia coronata* on the growth and reproductive capacity of *Avena barbata* and *A. fatua*. Within each group of figures (one species, one column) those values with common postscript letters do not differ significantly; all others differ at $P < 0.005$ except the time to anthesis for the early and late infections of *A. fatua* where $P < 0.01$

Treatment	Time to anthesis (days)	Dry weight (g)	Number of fertile tillers	Average weight per seed (mg)
Avena barbata				
Early infection	104·0[A]	5·7[A]	3·3[A]	0·105[A]
Late infection	106·8[A]	10·6[B]	5·3[B]	0·131[B]
Control	104·4[A]	17·4[C]	7·3[C]	0·137[B]
Avena fatua				
Early infection	131·3[A]	9·2[A]	5·3[A]	0·199[A]
Late infection	126·2[B]	19·4[B]	9·5[B]	0·267[B]
Control	124·9[B]	22·1[B]	9·5[B]	0·270[B]

sets of virulence genes but also, though to a lesser extent, those individuals possessing no or inappropriate qualitative resistance genes. Such protection of susceptible plants is achieved through a reduction in the amount of inoculum available for dispersal in the stand, through increased distances between susceptible plants and through the trapping of inoculum by resistant plants (Burdon 1978). The presence of quantitative resistance is also extremely important. It confers further protection for individuals possessing no qualitative resistance, provides the population as a whole with a means of buffering the worst effects of a pathogen in the event of the appearance of new virulence races (Burdon & Marshall 1981) and lessens the chance of the appearance of new virulence genes by reducing selective pressures on the pathogen population (Marshall & Burdon, unpublished data).

At present the importance of these different mechanisms in protecting plant populations can only be judged from a handful of host-pathogen combinations. Clearly, however, the contribution of each component is likely to vary from species to species and environment to environment. While I have touched on only one aspect of the interaction between pathogen activity and temporal diversity, at least part of the stability of natural plant-pathogen interactions is derived from this source. Uneven age distributions, differences in rates of development and changes in the susceptibility of plant parts with increasing age and from one stage of the life cycle to another may all act as stabilizing influences (Schmidt 1978).

PATHOGENS AND THE SPECIES DIVERSITY OF PLANT COMMUNITIES

The basic concepts concerning the interaction between fungal pathogens and their hosts apply also in plant communities, although the presence of additional species adds the further dimension of inter-specific competition to the host-pathogen interactions.

In healthy species stands, competitive interactions between component species are often complex, leading to either dominance by the more aggressive species or the development and maintenance of mechanisms which minimize niche overlap. In environments in which disease occurs these interactions are further complicated by those occurring between individual plants and their pathogens and between healthy and diseased plants. The outcome of these interactions may have a considerable impact on the aggressiveness of host species and hence on the final outcome of inter-specific competition (Burdon & Chilvers 1977b). Moreover, as the rate of increase of disease tends to be greater at high than at low plant density (Berger 1973; Burdon & Chilvers 1975, 1976a, unpublished data), the relative impact of pathogens in mixed stands is likely to be frequency dependent.

Theoretical considerations

The possible effect of pathogens on the species diversity of a community can be illustrated by developing a simple model based on the replacement series approach of de Wit (1960). In this model if we start initially with a simple competitive situation between two distinct plant species (A and B) which have the same environmental requirements and in which species A is more aggressive than species B, then in a disease-free environment the more aggressive species A will eventually exclude species B through a simple monopolization of resources (Fig. 1(a) and (d)).

If we now add to the model a host-specific pathogen capable of attacking species A alone, the outcome of the interaction may alter dramatically depending on the effect the pathogen has on the host plant. These possible effects may be grouped into three categories. Firstly, the impact of the pathogen may be minimal, having little effect on the growth of species A in pure stand or on its aggressiveness in mixtures. For such host-pathogen combinations the dominance of species A will continue and the replacement curves will remain the same as those occurring in disease-free situations (Fig. 1(a) and (d)).

Alternatively the pathogen may continue to have little effect on the yield of its host when grown in monoculture, but may have a marked effect on its ability to compete against species B. The general nature of the outcome of such an interaction is shown in the replacement line for species A in Fig. 1(b) and (e). When species A is present at high frequency (high proportion) in a mixture, conditions for the rapid increase and spread of disease from plant to plant are maximized, disease impact is severe and the competitive ability of A is strongly depressed. This results in the concave-shaped replacement line at high frequencies of A. On the other hand, when A is present at low frequencies, individual plants are well spaced and interspersed with non-susceptible individuals of species B, the spread of the pathogen from one individual to another is reduced and, hence, the rate of disease increase lowered. The impact of the pathogen on A is less severe than at high frequencies and the competitive advantage of A, while being reduced, still remains positive relative to species B. This gives a convex-shaped replacement line at low frequencies (Fig. 1(b)).

Finally, in the third possible interaction, the disease substantially reduces the growth of species A in monoculture and both its growth and competitive ability in mixture. In this situation a double inflected curve of the type obtained in Fig. 1(b) would again be expected, although the position of the inflexions is likely to be markedly different (Fig. 1(c)).

Of the three models described above, those illustrated in Fig. 1(b) and (c) are both potentially stabilizing. The pathogen, by preventing monopolization of environmental

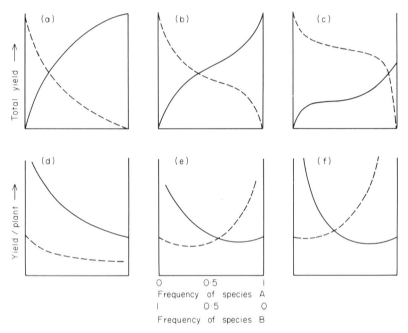

FIG. 1. Results of hypothetical replacement series experiments involving competition between two species having very similar resource requirements but different competitive abilities. The total number of plants per unit ground area is always the same but the frequencies (proportions) of A and B are varied. Species A is a stronger competitor than species B. (a)–(c) Total yield/unit area, (d)–(f) yield/plant. (a) and (d) Competition between species A (solid line) and species B (broken line) in the absence of disease. (b) & (e) and (c) & (f) Two potential outcomes of competition between species A and B in the presence of a pathogen capable of attacking species A alone. For further details see text.

resources by the more aggressive species, is responsible for the establishment of a dynamic equilibrium which may lead to an increase in community diversity.

If we now make these models more realistic by adding a set of host-specific pathogens capable of attacking species B and acknowledge the fact that 'real' species are genetically heterogeneous, the interactions between the two plant species will become even more complex but, at the same time, greater stability is likely to accrue. Moreover, in the longer term the complex of interactions between the two plant species and their respective pathogens is likely to result in the creation of further partly vacated niches as each set of pathogens prevent their hosts from taking full advantage of the reduced aggressiveness of their competitors. Ultimately this may lead to invasion and colonization by further species and their parasites.

A number of other models detailing the changes likely to result from interactions of different plant species in environments in which disease or parasites occur have been advanced (Janzen 1970; Chilvers & Brittain 1972; Roughgarden & Feldman 1975). Although these models all differ in the viewpoint from which the problem is approached and the mode of analysis used, they all confirm the strength of the basic concept. However, as before the question is whether such interactions occur in the real world.

Practical evidence

Considerable circumstantial evidence supporting the view that diseases play a role in determining the diversity of natural plant communities, or conversely that it is, at least in part, the diversity of natural communities which prevents disease epidemics from becoming commonplace, may be gleaned from comparisons of the frequency and severity of epidemics in agricultural cropping systems, managed semi-natural systems and undisturbed communities. Such comparisons show that disease epidemics are, in general, far more frequent and severe in agricultural systems (particularly those in which genetically uniform varieties are grown over large areas) than in managed semi-natural ones, which in turn, are more prone to damaging epidemics than are undisturbed plant communities (Browning 1974; Burdon & Shattock 1980). Some of these differences may reasonably be attributed to our greater awareness of disease in agricultural crops, but such comparisons still strongly support the proposed hypothesis.

The increasing use of mixtures of varieties or isolines of the one species as a means of disease control in agriculture is producing a wealth of factual data relevant to the whole question of the interaction between the diversity of a community and its potential susceptibility to pathogen attack. A number of studies have shown the success of such mixtures in reducing the rate of increase of epidemics (Leonard 1969; Burdon & Chilvers 1977a; Wolfe & Barrett 1977; Burdon & Whitbread 1979). Moreover, in at least one case (Wolfe & Minchin 1977) this reduction in disease levels at the vegetative phase has carried over into final yields with mixtures performing better than the mean of their component lines. In other cases (Browning & Frey 1969) such mixtures have produced a greater stability of grain yield but complete comparisons with monocultures of the component lines have not been carried out.

In a glasshouse experiment investigating competitive interactions between barley and wheat, Burdon & Chilvers (1977b) found that in situations in which the barley pathogen *Erysiphe graminis* f. sp. *hordei* was present, the competitiveness of the barley was significantly reduced while that of the wheat increased. Frequency dependent changes in competitive ability were not detected. Similar changes in competitive ability have been recorded in a field study of the effect of crown rust on the yield of ryegrass in mixed ryegrass–white clover swards. Here, over a 5-week period, crown rust infection depressed ryegrass yields by 84% while the unaffected white clover component of the sward responded to the weaker growth of ryegrass by an increase in yield of 87% (Latch & Lancashire 1970).

Experiments of this sort have not been carried out in more complex natural plant communities. In such communities the best hope of observing the operation of mechanisms such as diseases which are important in maintaining a stable equilibrium is to disturb the existing community balance through deliberate removals, additions or exclusions. Despite examples set in both insect-plant (e.g. Cantlon 1969) and plant-plant (see Harper 1977) studies this approach has not been used in the investigation of pathogen-plant interactions at a community level. The closest to such an approach comes in some of the attempts that have been made to control exotic weedy species with parasites from the plant's native range.

The use of fungal pathogens has lagged behind the use of insects in biological control; although a number of attempts have been made to use fungal pathogens specifically (for example, to control *Rubus* species with *Phragmidium violaceum* (Oehrens 1977)) and evidence has accumulated that fungal pathogens have contributed to the control achieved by insects (for example, *Cercospora eupatorii* in *Eupatorium adenophorum* control (Harley 1973)), it is only in the use of the rust fungus *Puccinia chondrillina* to control skeleton weed (*Chondrilla juncea*) that complete documentation has been achieved. Skeleton weed is an apomictic species, numerous different forms of which occur in the Mediterranean region where stand densities average less than 10 plants/m² (Wapshere *et al.* 1974; Wapshere, Caresche & Hasan 1976). By way of contrast, in Australia only three forms of the weed have been detected; prior to the release of *P. chondrillina* in 1971 one of these (Form A) was widespread throughout much of New South Wales, Victoria and South Australia, while the other two forms (B & C) were restricted to a small area in New South Wales (Hull & Groves 1973). In 1971 a single race of *P. chondrillina* capable of attacking the widespread form A of skeleton weed was released, rapidly became established (Cullen, Kable & Catt 1973) and has been responsible for reducing densities of this form from 200 to 10 plants/m² (Cullen & Groves 1977).

Further recent work on skeleton weed control has added to our knowledge of this interaction (Burdon, Groves & Cullen, in press). During the last decade there have been considerable changes in the distribution and abundance of the two more restricted forms. At sites at which, prior to the rust release of 1971, all three forms occurred together, there has been a change in the relative frequencies of the three forms with those of forms B and C increasing markedly while that of form A has decreased greatly. Similarly the geographic distribution of the previously restricted forms has increased considerably with form C, in particular, now occurring throughout much of the range of form A. Though the ranges of forms B and C were slowly expanding prior to the release of *P. chondrillina*, it is clear that the success of this pathogen in reducing the vigour and competitive ability of its host (form A) has been a major factor in the rapid expansion of the geographic distribution and abundance of the other two forms of the weed. Again this clearly demonstrates how a loss in competitive ability may lead to considerable floristic changes. In this particular case other resistant forms of the same species took advantage of the pathogen-induced changes. If these forms had not been present other species would almost certainly have occupied the partly vacated niches. This interaction also illustrates one of the ways in which pathogens may affect plant distribution. Others are now considered more fully.

PATHOGENS AND PLANT DISTRIBUTION

Surveys of the pattern of resistance of a number of different populations of one species or several closely related species have commonly found that populations in some areas show abnormally high levels of resistance when compared to the species as a whole. Often these geographic differences are correlated with differences in environmental conditions, with resistant populations being found in areas which particularly favour the pathogen. For example, Qualset (1975) found that the frequency of genotypes resistant to barley yellow dwarf virus was far greater in populations growing at high, than at low altitudes, and

Burdon & Marshall (1981) have shown that of four species of *Glycine* in Australia, those with a more northerly distribution, where conditions are optimal for the spread and increase of disease caused by *Phakopsora pachyrhizi*, show greater resistance than do species restricted to areas less suitable to the pathogen.

With such clear evidence of differences in the geographic distribution of resistance it is not hard to accept that in some circumstances environmental conditions may be so favourable to the pathogen that host and pathogen are restricted to areas where they are in more even balance. Such a possibility seems a reasonable explanation for the natural restriction of European larch to the Alps. In this area this species forms extensive natural stands in which the pathogen *Trichoscyphella willkommii* is of only minor importance. When larch is grown in similarly extensive stands in lowland Europe, however, this pathogen becomes extremely damaging (Large 1940).

Soil-borne pathogens are often even more sensitive to changes in environmental conditions than are airborne ones and it is possible to imagine situations where interactions between such pathogens and their hosts might result in the occurrence of much smaller refuge areas intimately intermingled with areas in which the pathogen is strongly favoured. Such a hypothesis has been advanced to explain the restriction of eucalypt species highly susceptible to attack by the pathogen *Phytophthora cinnamomi* to dry ridge top sites in southeastern Australia (Heather, Pratt & Shepherd 1975; Burdon & Shattock 1980).

Finally diseases may even play a role in the distribution of species within physically homogeneous areas. In a plant community, reductions in plant densities in response to pathogen pressure may be achieved by the selective removal of individuals so as to produce randomly or regularly distributed plants. Equally, however, similar density reductions may be achieved through the creation of localized patches of high density dispersed against a background of low or zero density. Such aggregated patterns of plant distribution are common in plant communities and may in fact contribute to disease control. In the only extensive experimental study of this possibility Burdon & Chilvers (1976b) showed that in clumped seedling stands the rate of increase of disease was, in general, slower than in unclumped, evenly spaced stands of the same overall density. This decline in the rate of increase, though not great, demonstrated the *possible* contribution of clumped distributions to disease control.

CONCLUSION AND FUTURE APPROACHES

The magnitude of response of a host plant population to a pathogen is a measure of the extent to which coevolution has occurred. As we have seen, a species may respond to selective pressure from a pathogen through the deployment of both qualitative and quantitative resistance, through increasing tolerance to pathogen attack, through disease avoidance either in time or space and through reductions in density and intermingling with individuals of other non-host species. Different host species and, in fact, different populations of the one species are likely to utilize these options differently, so that in many undisturbed, coevolved systems damage is little evident and the pathogen functions as part of the general homeostatic mechanism maintaining genetic, spatial and demographic variability. It is timely to stress that day-to-day levels of disease incidence are not a good criterion by which to judge whether a pathogen is an important factor in the demography of a plant species. As Haldane & Jayakar (1962) have shown, once a resistant/susceptible polymorphism has been established in a population, even if susceptible individuals have

5–10% greater fitness in disease-free situations, the sporadic occurrence of an epidemic which kills all susceptible plants only once every 10 yr is quite sufficient to maintain a dynamic polymorphism in the population.

Clearly our understanding of the ecological role of plant pathogens is at present little more than a skeletal framework of postulated interactions. In the future this skeleton has to be 'fleshed out' and restructured so that plant-pathogen interactions may take their rightful position in any ecological consideration of the factors affecting plant populations and communities. There is no doubt that one of the most interesting ecological approaches that could be used is that of monitoring changes occurring as a result of the deliberate removal or addition of selected species, the deliberate creation of dense patches of potential hosts and the deliberate exclusion of pathogens from existing plant communities. Moreover, a demographic approach to studying the life cycles of individual plant species will also provide an opportunity for investigating the relative importance of the impact of diseases at different stages in the life cycle. Because of the lack of suitable examples, I have not considered here the impact of disease on the germination and establishment phase of the life cycle. However, the impact of pathogens at this stage is likely to be quite different to the impact of pathogens on adult plants (Burdon & Shattock 1980) and may in fact lead to faster genetic changes in the host population.

As Harper (1977) has pointed out, many of the ideas and techniques of plant pathology are also relevant to this area of population biology and should not be ignored. The continuing use of multilines and varietal mixtures for agricultural crops provides opportunities to explore competitive interactions in the presence of disease on a large scale, thus hopefully overcoming the interference problems that plague small plot experiments. Equally, each new biological control programme incorporating fungal pathogens represents a 'one-off' field experiment in which many of the ideas discussed here can be examined.

Finally we should continue to speculate. While the suggestion that parasite pressure in general might be important in influencing the recombination system of many plant species (Levin 1975) is basically untestable, such speculation is valuable in drawing evidence together and in alerting us to the potential far-reaching ramifications of disease in plant communities.

ACKNOWLEDGMENTS

The helpful comments and suggestions of Drs Adrian Gibbs, W. A. Heather, D. R. Marshall and A. O. Nicholls are gratefully acknowledged.

REFERENCES

Barrett, J. A. (1978). A model of epidemic development in variety mixtures. *Plant Disease Epidemiology* (Ed. by P. R. Scott & A. W. Bainbridge), pp. 129–137. Blackwell Scientific Publications, Oxford.

Berger, R. D. (1973). Infection rates of *Cercospora apii* in mixed populations of susceptible and tolerant celery. *Phytopathology*, **63**, 535–537.

Browning, J. A. (1974). Relevance of knowledge about natural ecosystems to development of pest management programs for agro-ecosystems. *Proceedings of the American Phytopathological Society*, **1**, 191–199.

Browning, J. A. & Frey, K. J. (1969). Multiline cultivars as a means of disease control. *Annual Review of Phytopathology*, **7**, 355–382.

Burdon, J. J. (1978). Mechanisms of disease control in heterogeneous plant populations—an ecologist's view. *Plant Disease Epidemiology* (Ed. by P. R. Scott & A. W. Bainbridge), pp. 193–200. Blackwell Scientific Publications, Oxford.

Burdon, J. J. (1980). Variations in disease resistance within a population of *Trifolium repens*. *Journal of Ecology*, **68**, 737–744.

Burdon, J. J. & Chilvers, G. A. (1975). Epidemiology of damping-off disease (*Pythium irregulare*) in relation to density of *Lepidium sativum* seedlings. *Annals of Applied Biology*, **81**, 135–143.

Burdon, J. J. & Chilvers, G. A. (1976a). Controlled environment experiments on epidemics of barley mildew in different density host stands. *Oecologia*, **26**, 61–72.

Burdon, J. J. & Chilvers, G. A. (1976b). The effect of clumped planting patterns on epidemics of damping-off disease in cress seedlings. *Oecologia*, **23**, 17–29.

Burdon, J. J. & Chilvers, G. A. (1977a). Controlled environment experiments on epidemics of barley mildew in different mixtures of barley and wheat. *Oecologia*, **28**, 141–146.

Burdon, J. J. & Cnilvers, G. A. (1977b). The effect of barley mildew on barley and wheat competition in mixtures. *Australian Journal of Botany*, **25**, 59–65.

Burdon, J. J., Groves, R. H. & Cullen, J. M. (1981). The impact of biological control on the distribution and abundance of *Chondrilla juncea* in southeastern Australia. *Journal of Applied Ecology*, **18**, 957–966.

Burdon, J. J. & Marshall, D. R. (1981). Inter- and intra-specific diversity in the disease response of four *Glycine* species to the leaf-rust fungus *Phakopsora pachyrhizi*. *Journal of Ecology*, **69**, 381–390.

Burdon, J. J. & Shattock, R. C. (1980). Disease in plant communities. *Applied Biology*, **5**, 145–219.

Burdon, J. J. & Whitbread, R. (1979). Rates of increase of barley mildew in mixed stands of barley and wheat. *Journal of Applied Ecology*, **16**, 253–258.

Cantlon, J. E. (1969). The stability of natural populations and their sensitivity to technology. *Brookhaven Symposia in Biology*, **22**, 197–203.

Chilvers, G. A. & Brittain, E. G. (1972). Plant competition mediated by host-specific parasites—a simple model. *Australian Journal of Biological Sciences*, **25**, 749–756.

Cullen, J. M. & Groves, R. H. (1977). The population biology of *Chondrilla juncea* L. in Australia. *Proceedings of the Ecological Society of Australia*, **10**, 121–134.

Cullen, J. M., Kable, P. F. & Catt, M. (1973). Epidemic spread of a rust imported for biological control. *Nature*, **244**, 462–464.

de Wit, C. T. (1960). On competition. *Verslag Landbouwkundige Onderzoekingen*, No. 66.

Dinoor, A. (1970). Sources of oat crown rust resistance in hexaploid and tetraploid wild oats in Israel. *Canadian Journal of Botany*, **48**, 153–161.

Dinoor, A. (1977). Oat crown rust resistance in Israel. *Annals of the New York Academy of Sciences*, **287**, 357–366.

Harlan, J. R. (1976). Diseases as a factor in plant evolution. *Annual Review of Phytopathology*, **14**, 31–51.

Haldane, J. B. S. & Jayakar, S. D. (1962). Polymorphism due to selection of varying direction. *Journal of Genetics*, **58**, 237–242.

Harley, K. L. S. (1973). Biological control of Central and South American weeds in Australia. *Proceedings of the II International Symposium on Biological Control of Weeds, Rome 1971*, pp. 4–10.

Harper, J. L. (1969). The role of predation in vegetational diversity. *Brookhaven Symposia in Biology*, **22**, 48–62.

Harper, J. L. (1977). *Population Biology of Plants*. Academic Press, London.

Heather, W. A., Pratt, B. H. & Shepherd, C. J. (1975). The impact of Pythiaceous fungi in Australian native forests. *Proceedings, Second International FAO/IUFRO World Technical Consultation on Forest Disease and Insects*. New Delhi, April 1975.

Hull, V. J. & Groves, R. H. (1973). Variation in *Chondrilla juncea* L. in south-eastern Australia. *Australian Journal of Botany*, **21**, 113–135.

Janzen, D. H. (1970). Herbivores and the number of tree species in tropical forests. *American Naturalist*, **104**, 501–528.

Large, E. C. (1940). *The Advance of the Fungi*. Jonathan Cape, London.

Latch, G. C. M. & Lancashire, J. A. (1970). The importance of some effects of fungal disease on pasture yield and composition. *Proceedings, XIth International Grassland Congress*, pp. 688–691.

Leonard, K. J. (1969). Factors affecting rates of stem rust increase in mixed plantings of susceptible and resistant oat varieties. *Phytopathology*, **59**, 1845–1850.

Leonard, K. J. (1977). Selection pressures and plant pathogens. *Annals of the New York Academy of Sciences*, **287**, 207–222.

Levin, D. A. (1975). Pest pressure and recombination systems in plants. *American Naturalist*, **109**, 437–451.

Marshall, D. R. (1977). The advantages and hazards of genetic homogeneity. *Annals of the New York Academy of Sciences*, **287**, 1–20.

Marshall, D. R. & Pryor, A. J. (1978). Multiline varieties and disease control. I. The 'dirty crop' approach with each component carrying a unique single resistance gene. *Theoretical and Applied Genetics*, **51**, 177–184.

Nelson, R. R. (1978). Genetics of horizontal resistance to plant diseases. *Annual Review of Phytopathology*, **16**, 359–378.

Oehrens, E. (1977). Biological control of blackberry through the introduction of the rust, *Phragmidium violaceum*, in Chile. *F.A.O. Plant Protection Bulletin*, **25**, 26–28.

Person, C. (1966). Genetic polymorphism in parasitic systems. *Nature*, **212**, 266–267.

Qualset, C. O. (1975). Sampling germplasm in a centre of diversity: an example of disease resistance in Ethiopian barley. *Crop Genetic Resources for Today and Tomorrow* (Ed. by O. H. Frankel & J. G. Hawkes), pp. 81–96. Cambridge University Press.

Roughgarden, J. & Feldman, M. (1975). Species packing and predation pressure. *Ecology*, **56**, 489–492.

Schmidt, R. A. (1978). Diseases in forest ecosystems: the importance of functional diversity. *Plant Disease: an Advanced Treatise*, Vol. 2 (Ed. by J. G. Horsfall & E. B. Cowling), pp. 287–315. Academic Press, New York.

Segal, A., Manisterski, J., Fischbeck, G. & Wahl, I. (1980). How plant populations defend themselves in natural ecosystems. *Plant Disease: an Advanced Treatise*, Vol. 5 (Ed. by J. G. Horsfall & E. B. Cowling), pp. 75–102. Academic Press, New York.

Stern, K. & Roche, L. (1974). *Genetics of Forest Ecosystems*. Springer-Verlag, Berlin.

Wahl, I. (1958). Studies on crown rust and stem rust of oats in Israel. *Bulletin of the Research Council of Israel, Section D*, **6**, 145–166.

Wahl, I., Eshed, N., Segal, A. & Sobel, Z. (1978). Significance of wild relatives of small grains and other wild grasses in cereal powdery mildews. *The Powdery Mildews* (Ed. by D. M. Spencer), pp. 83–100. Academic Press, New York.

Wapshere, A. J., Caresche, L. & Hasan, S. (1976). The ecology of *Chondrilla juncea* in the eastern Mediterranean. *Journal of Applied Ecology*, **13**, 545–553.

Wapshere, A. J., Hasan, S., Wahba, W. K. & Caresche, L. (1974). The ecology of *Chondrilla juncea* in the western Mediterranean. *Journal of Applied Ecology*, **11**, 783–799.

Wolfe, M. S. & Barrett, J. A. (1977). Population genetics of powdery mildew epidemics. *Annals of the New York Academy of Sciences*, **287**, 151–163.

Wolfe, M. S. & Minchin, P. N. (1977). Effects of variety mixtures. *Annual Report of the Plant Breeding Institute, Cambridge* 1976, pp. 112–114.

ANTAGONISMS IN THE REGENERATION OF *EUCALYPTUS REGNANS* IN THE MATURE FOREST

D. H. ASHTON and E. J. WILLIS*

School of Botany, University of Melbourne, Parkville, Victoria 3052, Australia

SUMMARY

Regeneration of *Eucalyptus regnans* in the mature forest usually fails in the absence of fire. Apart from the normal environmental hazards of fungal attack, drought and heat stress, seed harvesting and marsupial browsing, there appear to be antagonistic soil factors. The mature trees may contribute to this by increasing the lipid content of the top soil or by the action of root exudates. Local peat accumulation at old tree-butts is very acid and very inhibitory to seedling growth. Soil disturbance and incubation at very warm temperatures leads to more favourable seedling growth with no symptoms of phosphorus deficiency. The rhizosphere fungus *Cylindrocarpon destructans* is common and yields a powerful toxin in culture. This species and several others show marked differences in abundance between healthy and unhealthy seedlings. It is suggested that combinations of rhizosphere species may be important in ameliorating any antagonistic effects. Changes in the microflora which are a consequence of fires may also prove to be one of the factors involved in the successful growth of *E. regnans* in very large gaps.

INTRODUCTION

One of the enigmatic problems in wet, undisturbed eucalypt forests in Australia is the lack of regeneration of the dominants in the absence of fire. Under annual rainfalls of 1100–2200 mm these forests grow rapidly to heights of 50–100 m and permit the development of a dense, shade-casting understorey of notophyllous trees and shrubs. In many places in the east of the continent rainforest encroaches beneath the canopy of giant emergents, where fires have not occurred for intervals of 50–200 yr (Jackson 1968).

The magnificent mature forest of *Eucalyptus regnans*† in the Wallaby Creek area of the Great Dividing Range, 65 km N.N.E. of Melbourne is one such forest where regeneration is notably absent. An even-aged forest arose after fires *c.* 1730, and to some extent again in 1851. The tall understorey of *Pomaderris aspera* (15–20 m) developed after surface fires in 1898. During 30 yr observation, the thinning-out of overstorey and understorey has continued steadily and the tree fern and ground fern strata have thickened commensurately. The only natural regeneration ever observed has occurred on leaning tree fern trunks or on the upthrown root mounds of giant trees. In no instance have such seedlings or saplings persisted for more than a few years. Regeneration in both scarified and undisturbed seed-beds failed under the full canopy of overstorey and understorey due to

* Present address: Melbourne State College, Parkville, Victoria 3052, Australia.
† Nomenclature of vascular plants follows Willis (1970, 1972).

0262–7027/82/0300–0113$02.00 © 1982 British Ecological Society

fungal attack and desiccation of weakened, shaded plants. In the better illumination beneath the mature canopy alone, growth continued through the juvenile stage but was decimated by winter fungal disease (Ashton & Macauley 1972). In moderately large gaps growth was satisfactory for 5 yr, after which time plants became suppressed and were replaced by the natural regeneration of the faster-growing understorey species, *Pomaderris aspera* and *Prostanthera lasianthos*. Under natural conditions very few of the current year's seed fall of *E. regnans* escaped harvesting by insects—principally the nocturnal ant, *Prolasius pallidus* (Ashton 1979). Seedlings surviving to the conspicuous 4-leaf stage were remorselessly browsed by wallabies (*Wallabia bicolor*), which eliminated all but the most protected plants. The light factor is critical for a heteroblastic species such as *E. regnans*. Seedlings in their first year are well-adapted to intercept overhead light; thereafter, as leaves become isobilateral and pendulous, their need for uninterrupted lateral light increases. Their light compensation point also increases with age, and thus their ability to outgrow the damaging attacks of insects and fungi under shaded conditions is progressively impaired (Ashton & Turner 1979). Seedlings germinating on soil under the combined shade of the overstorey and understorey (4–8% daylight) will develop only as far as the 2-leaf stage and do not survive more than 1 yr.

Seedlings planted at the 6–8-leaf stage under these canopy conditions will develop intermediate foliage and grow and etiolate to a height of 30–40 cm before succumbing after 2–3 yr. Improved growth of *E. regnans* seedlings occurs if the roots of *E. regnans* and *Pomaderris aspera* trees are severed to a depth of 30 cm. Such seedlings may live 2 yr and develop their full juvenile condition of 10–12 opposite leaves. Similar growth occurs on rotten logs and on tree fern trunks. Growth and form under this canopy is compatible with that under 5–10% daylight in the glasshouse. Under the canopy of *E. regnans* alone, seedlings develop fairly rapidly to the intermediate stage with large isobilateral leaves. However, after 5–6 yr these too become suppressed, and after 10–12 yr they are replaced by the more vigorous understorey species *Pomaderris aspera*, *Prostanthera lasianthos* and *Acacia dealbata*. By contrast, regeneration initiated under mature canopy by surface fires will persist for 30–40 yr as suppressed saplings and poles, 10–20 m high. In the absence of mature trees over large areas of 1 ha or more, the growth of *E. regnans* regeneration is normal, and understorey species never attain dominance.

The extreme suppression of very young *E. regnans* seedlings in natural soil in the field under moderate shade of the understorey trees, and the progressive suppression of advanced seedlings and saplings under the light shade of the mature canopy suggest that inhibitory factors may interact, probably synergistically, with the normal physical, chemical and biotic stresses of the environment.

Many antagonistic effects are possible in the mature forest environment, and research in Australia has been pioneered by Florence & Crocker (1962) in wet sclerophyll forests of *E. pilularis* in New South Wales. Beginning in 1957, aspects of allelopathic influence were investigated by one of us (D.H.A.) in the mature forests of *E. regnans* at Wallaby Creek, and from 1972 to 1980 in pole and mature forests at Toolangi and Warburton, 60–80 km to the east and north-east of Melbourne (Willis 1980).

Some of the questions posed by Dr A. S. Watt during his stay in Melbourne in 1950 are still being pondered, and the ramifications of the subsequent research have been very largely due to his initial guidelines and his deep continuing interest.

DIRECT ALLELOPATHIC EFFECTS OF THE MATURE TREE

Foliar leachates

Del Moral & Muller (1969) found that the natural foliar leachates (fog drip) of *E. globulus*, grown as an exotic in California, inhibited the germination and growth of understorey herbs. Work with *E. camaldulensis* in California established that both water-soluble and volatile compounds could inhibit not only herb, but also *E. camaldulensis* seedling development (del Moral & Muller 1970). More recently, Al-Mousawi & Al-Naib (1975, 1976) and Al-Maib & Al-Mousawi (1976) suggested that similar mechanisms may inhibit herb growth beneath the canopy of *E. microtheca* in Iraq. While the above studies were performed where *Eucalyptus* was grown as an exotic, del Moral, Willis & Ashton (1978) found that foliar leachates of *E. baxteri*, growing in its native habitat in Wilson's Promontory, Victoria, may suppress the growth of understorey shrubs such as *Casuarina pusilla*.

The foregoing, particularly in conjunction with the known phenolic content of eucalypt foliage (Hillis 1966, 1967) and the high rainfall characteristic of *E. regnans* forests, suggested that foliar leachates from the canopy of *E. regnans* may provide an allelopathic mechanism of seedling suppression. Extensive bioassays of spray-derived leachates from both intact foliage and litter of *E. regnans* have shown no inhibitory activity, except when such leachates were so highly concentrated (foliar ×50, litter ×10) that osmotic inhibition occurred (Willis 1980). Extracts of at least two understorey shrubs, *Cassinia aculeata* and *Pittosporum undulatum*, did show allelopathic potential.

In view of the relatively wet environment in which *E. regnans* occurs, it is unlikely that volatile compounds would provide an effective mechanism of toxin release (Muller 1970). In fact, the essential oil composition of leaves of *E. regnans* is uncharacteristic of *Eucalyptus*, and is dominated by a waxy sesquiterpene, eudesmol (Baker & Smith 1920). Aqueous extracts of the distilled oil can be shown to be highly toxic in bioassay; however, natural foliar leachates contain no terpenoids, and thus a mode of release is not known. In view of the extensive insect attack on *E. regnans* foliage by members of the Scarabaediae, Chrysamelidae and Phasmatidae, the rain of insect frass may provide an avenue of terpene release, as found by Trenbath & Fox (1976) with *E. bicostata*.

Soil lipids

In testing soils from *E. regnans* forests for phenolic compounds, it was observed that topsoil was particularly rich in lipids. These compounds constitute a diverse group of organic compounds, including waxes, resins, long-chain fatty acids, alcohols, ketones, and others, whose main common characteristic is their solubility in non-polar solvents. The effects of naturally-occurring lipid compounds in soil have been the subject of speculation since the work of Schreiner, Reed & Skinner (1907). Although it has been suggested that soil lipids may act directly as inhibitors (Häsler & Wanner 1977), it is likely that, in most systems, soil lipids act in modifying nutrient availability (Romashkevich 1964) or in altering the microbiological status of the soil (Otroshchenko *et al.* 1979).

The topsoil (0–10 cm) from pole stage *E. regnans* forest at Toolangi was found to be rich in lipid material, and when extracted with 2:1 benzene-ethanol for 8 h in a Soxhlet apparatus the soil yielded up to 3·5% lipid (on a dry weight basis). Newly germinated seedlings of *E. viminalis* were grown in Petri-dishes on two layers of filter paper in which

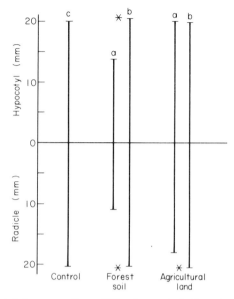

FIG. 1. Growth of cotyledonary seedlings of *Eucalyptus viminalis* under the influence of lipids extracted from krasnozem top soil from *E. regnans* forest or agricultural (potato) land. a, seedling in direct contact with lipid-impregnated filter paper; b, seedling protected from such paper by a fresh filter paper; c, control with two fresh filter papers. * difference between a and b statistically significant ($P \leqslant 0.05$).

either the upper or lower paper was impregnated with soil lipid. Controls consisted of clean filter paper only. All treatments were moistened with distilled water. Severe inhibition of extension growth of both hypocotyl and radicle occurred where the plant was in direct contact with lipid (Fig. 1). Root hair development was suppressed. Subsequent experiments demonstrated that seedling growth in soil was enhanced if lipids were dissolved from the soil; however, the solvent also markedly affected soil structure and, almost certainly, the microflora. Comparative gas-liquid chromatography revealed that the lipids are largely of leaf cuticular origin (Willis 1980). *Eucalyptus viminalis* gave the same response as *E. regnans* to lipid contact and was used in the bioassay because of its superior germination characteristics.

The accumulation of lipid material in the soil of forests of *E. regnans* is likely to vary with age of the stand, site, and drainage conditions. In the mature forest at Wallaby Creek the lipid content of the soil was only 0·4–0·6% and diminished by 25% when vigorous seedlings of *E. regnans* were grown in the soil for 5–6 months. However, the initial growth of *E. regnans* on freshly collected soil is poor (see later section) but whether this temporary retardation coincides with detoxification or microbial alteration of lipid compounds remains unknown.

The effect of old root systems

Foresters have long appreciated the significance of root competition under the shade of tree canopies (Toomey & Korstian 1947). The precise nature of such effects, however, is not easily elucidated. Early work attributed depressed growth to competition for water and nutrients. Work by Robinson (1971) has indicated the complexity of inter-mycorrhizal antagonism.

Eucalyptus regnans suffers undue suppression if growing in unburnt soil under mature canopy of the forest. In 1955 Professor F. W. Went (personal communication) suggested that exudations from the living root systems of mature *E. regnans* may inhibit its juvenile stages. In 1957 a pilot trial was set up in triplicate in the glasshouse, where cotyledonary seedlings were planted into krasnozem soil in buckets of 9 litre capacity. In one set of buckets large plants of *E. regnans* or *Pomaderris aspera*, 1·5–2·0 m tall, had been established. Nutrient solutions were added to half of each treatment every 2–3 weeks, and watering was carried out with careful attention to the needs of the young seedlings. After 9 months, the average height growth of the control plants with fertilizer was 71·2 cm, without fertilizer 34·8 cm. By contrast, the growth of seedlings associated with the large plants was 4·6 and 1·6 cm respectively. Inhibition was thus associated with both *E. regnans* and *P. aspera*.

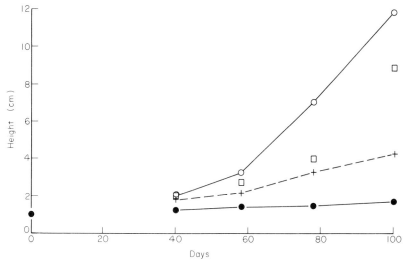

FIG. 2. The height-growth of seedlings in relation to root contact of large plants under glasshouse conditions. ○, sealed pots; □, defectively sealed pots; +, porous pot; ●, open-rooted.

In 1961 this experiment was enlarged and replicated ten-fold. Cotyledonary seedlings were grown (1) open-rooted in buckets as previously, (2) in porous terracotta pots 7·5 cm in diameter which had been counter-sunk into the soil in buckets or (3) in similar pots sealed off from contact with outside roots by thick plastic sheeting and drained to the outside of the bucket.

The plants were well watered and given frequent applications of complete fertilizer. After 100 days the open-rooted plants remained in the cotyledonary stage, those in the porous pots had developed 2 leaves, and those in the sealed pots had reached the 6-leaf stage (Fig. 2). Three of the latter treatment which were poorer than the remainder were found to have had defective plastic wrapping, and the surface of the pots was enmeshed by mycorrhizal roots from the large plant. There seemed to be a *prima facie* case supporting the suggestion made by Went in 1955, although the high CO_2 concentration in soil of the large seedling may have exaggerated this effect.

In 1962, leachates from buckets containing large plants and from control Krasnozem soil were watered on to 2-leaf stage seedlings in the glasshouse every 2 days for 2 months. The results in Table 1 indicate that height and dry weight of seedlings receiving leachate

TABLE 1. Height and dry weight of *Eucalyptus regnans* seedlings watered with leachates from very large potted plants of *E. regnans* in the glasshouse under conditions of 60% daylight. Large plants, although somewhat pot-bound, were in active growth. Significance tested by Student's *t* statistic; twenty replicates

	Dry weight (g/plant)	Height (cm)
Normal concentration		
Leachate from large plant	0·08	3·7
Leachate from forest soil	0·18 $*P \leqslant 0.02$	5·5 $*P \leqslant 0.05$
Concentration × 10		
Leachate from large plant	0·38	8·1
Leachate from forest soil	0·94 $*P \leqslant 0.01$	14·4 $*P \leqslant 0.05$
Tap water only	0·19	5·5
Tap water on to forest litter mulch	0·19 NS	6·0 NS

	Large seedling (actively growing)	Forest soil
CO_2 concentration‡		
At 2·5 cm depth	3·4%	0·82%
5·0 cm depth	4·28%	1·54%
10·0 cm depth	5·84%	1·41%

‡ Collected with syringe, measured by IRGA.

from large potted plants were significantly less than those receiving either tap water or leachate from forest top-soil. The average CO_2 content of soil air was higher in large seedling pots than in forest top-soil.

To test the hypothesis that living roots *per se* affect the growth of seedlings, experiments were carried out in the mature forest at Wallaby Creek. In 1962, 2-leaf stage seedlings in Krasnozem soil were planted open-rooted in alternation with those countersunk in sealed 10-cm diameter pots. All plants were manured with blood and bone at the rate of 14

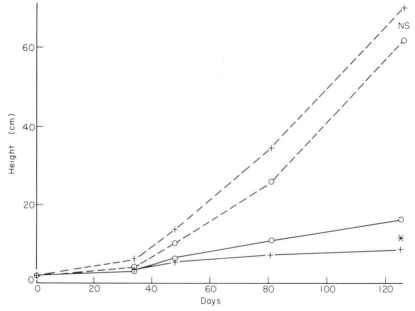

FIG. 3. Height growth rates of potted and open-rooted seedlings in summer in the mature forest. O, sealed pots; +, open-rooted; – – – –, half canopy; ———, full canopy. * statistically significant at $P \leqslant 0.001$.

Table 2. Dry weight (g) of *Eucalyptus regnans* seedlings after growth in mature forest under Full Canopy for 23 weeks, 1964–65. Statistical significance tested by Student's *t* statistic, NS = not significant

	+ nutrient solution	+ water only	Statistical significance (nutrients/water)
Sealed pot	7·65	7·61	NS
Open-rooted	5·81	5·11	NS
Statistical significance (sealed/open-rooted)	$P \leqslant 0.05$	$P \leqslant 0.02$	

g/plant, and watered each week at a rate equivalent to 9 mm of rain. The average dry weight of fine roots (<2 mm diameter) m^{-2} in the top 15 cm of soil in the Half Canopy and Full Canopy sites were 11·7 g (*E. regnans*) and 68·4 g (*E. regnans* and *P. aspera*) respectively. After 4 months under Full Canopy (overstorey and understorey) conditions the height growth by plants in pots was significantly greater than those open-rooted in the soil (Fig. 3). In the Half Canopy (overstorey only) conditions, no significant difference occurred over this period. This latter result may be anticipated because of the low root density of mature trees, which were 15–20 m distant. Suppression of seedlings in these canopy conditions was not evident in earlier trials until the second or third year.

Two further experiments were carried out at the Full Canopy site during summer. In 1964 seedlings were planted at the 6–8 leaf stage; in 1967 seedlings planted at the 2, 4 and 6–8 leaf stage were compared. As in 1962, open-rooted and sealed pots were compared, but here half of the pots in each treatment received NPK fertilizer, the others tap water only. In both experiments there was, as before, significantly enhanced growth when the plants were potted, though the youngest seedlings did not show this during this period (Table 2, Fig. 4). The addition of fertilizer did not remove this enhancement, though it may

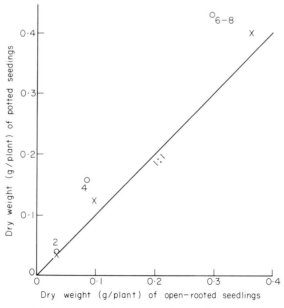

Fig. 4. Dry weight (g) of seedlings at different stages in potted or open soil with (×) or without (O) complete nutrients (N as NH_4^+) beneath the full canopy of the *Pomaderris aspera* understorey. Figures indicate number of leaves per seedling at planting.

have reduced it. These experiments serve to emphasize the need to assess the effect of the plant and age on the response to the environment. However, it seems likely that some inhibitory factor may be produced from the soil or the heavily ectomycorrhizal roots of both *E. regnans* and *P. aspera* which causes diminished growth in very young, poorly mycorrhizal seedlings in shaded environments (Ashton 1976). The severing of competing tree roots improved, to some extent, the growth and survival of seedlings. However, no inhibition has been obtained by the experimental addition of severed roots to soil, either in the glasshouse or laboratory bioassay.

These observations may be resolved if it is postulated that inhibitory substances are a consequence of the living root system. What they may be or how long they persist is not known.

The effects of mature trees on the soil

It has been long known that different tree species produce local effects on soil (Müller 1887; Watt 1961; Richards 1952). In California, Zinke (1962) showed that the effects of old trees on the soil were zoned according to the predominance of bark litter, leaf and twig litter and throughfall.

Striking 'sugar-sand' podzolization occurs beneath large *Agathis australis* in New Zealand (Taylor & Pohlen 1970), and large mounds of peat occur beneath the crown projection of old *Fitzroya cupressoides* in Chile (Veblen & Ashton 1982). Marked decreases in pH around *Eucalyptus pilularis* and *E. microcorys* trees in Queensland have been found by Ward (1979) and the large accumulation of pure humus around *E. regnans* trees in Tasmania has been cogently argued by Cremer (1962) to be a major cause of tree death by girdling during ground fires.

At Wallaby Creek the mature forest returns a mean annual litter fall of 8·08 t/ha, of which approximately 50% is leaf material, 26% wood and 18% bark. Almost all of the leaf material disintegrates and decomposes in one year, thus ensuring a rapid recycling of nutrients (Ashton 1975). The organic matter content in the upper 10–15 cm is about 12–18% and the biological activity of the soil, as judged from the rate of CO_2 production, is high (Ellis 1969). Around each tree, very heavy accumulations of bark occur from both the deciduous bark strips of the trunk and the bryophyte mats, and subfibrous bark of the butt. As a result, friable amorphous peat has developed to a depth of 10–20 cm over a

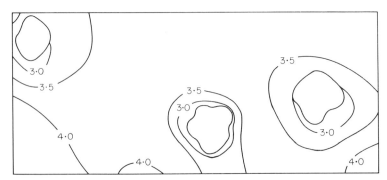

FIG. 5. Isolines of equal pH around three mature 250 yr-old trees of *Eucalyptus regnans* under a mature understorey of *Pomaderris aspera*. Measured using a glass electrode directly into the top soil (0–10 cm). Values are 0·5 units higher when tested in the laboratory using in a soil:water ratio of 1:2·5. Plot was 20 × 8·5 m.

zone up to 20–40 cm from the trunk. The extreme acidity of these zones is shown in Fig. 5. The development of any species but ferns in this material is rare. *E. regnans* seedlings which occasionally establish in these moist microsites become stunted and do not persist more than 1–3 yr.

A glasshouse bioassay carried out on the upper 10–12 cm of butt soil and inter-tree soil from Wallaby Creek revealed severe stunting of *E. regnans* seedlings for up to 6 months (Table 3). During this time the peat samples exuded a dark brown solution rich in tannins. After 9 months, growth and appearance of seedlings in this medium was normal, although pH of the butt soil remained one order of magnitude lower (3·90) than that of the inter-tree top-soil. The nutrient content of the butt peats on a dry weight basis was surprisingly similar to that of the inter-tree top-soil, with the exception of the C/N ratio. The

TABLE 3. Chemical analysis and bioassay of tree-butt peat and inter-tree top-soil (0–10 cm) under mature understorey of *Pomaderris aspera*

Chemical analysis (Food Laboratories, Carlton, Victoria) on a dry-weight basis

	Tree-butt peat	Inter-tree top-soil
Organic matter (%)	45·6	18·0
Carbon (%)	26·5	10·6
Nitrogen (%)	1·15	0·68
C/N	23·0	15·6
'Available' P (μg g^{-1}) (NaHCO$_3$ extract)	12	6
Exchangeable Ca (μg g^{-1})	1300	1600
Exchangeable Mg (μg g^{-1})	380	320
Exchangeable K (μg g^{-1})	52	42
Apparent specific gravity (g cm^{-3})	0·17	0·62

Bioassay of *E. regnans* seedlings in glasshouse

	At 4 months		At 6 months		At 12 months	
	Butt peat	Top-soil	Butt peat	Top-soil	Butt peat	Top-soil
Dry weight (g)	0·31 *	5·02	–	–	10·7 *	22·0
Height (cm)	2·9 *	6·2	4·2 *	29·5	45·0	49·5
pH (1:5 dilution)	3·5	4·5	–	–	3·9	4·7

* Difference statistically significant at $P \leqslant 0·001$.

mycorrhizae of *E. regnans* trees which ramify through the peat zone are distinct from the normal types, and are typically dark brown with white sheaths.

The local development of such inhibitory soil from the degradation of specific litter types indicates the extent to which allelopathic response is possible in this forest.

THE EFFECTS OF MICROBIAL INTERACTIONS

Nitrification

Rice & Pancholy (1972) suggested that, during the course of succession, soil nitrification should decrease, due largely to inhibition by the dominant vegetation.

In a preliminary study at Wallaby Creek, Victoria, top-soil from mature *E. regnans* forest was found to have an ammonium:nitrate ratio of 6, whereas soil from corresponding pole-stage forest had a value of 2. In a more detailed study in a wetter forest at Toolangi, Victoria, samples taken bi-monthly over a 1-yr period showed that although *E. regnans* soils were nitrifying they showed considerable seasonal variation between different aged stands. However, when top-soil was removed from spar-stage *E. regnans* forest in spring a forty-fold increase in nitrate occurred after 8 day's incubation at room temperature. This suggests that if inhibitors are present they are short-lived.

Seedlings of *E. regnans* showed no marked preference for ammonium or nitrate as a nitrogen source. Thus the nitrogen status of the mature forest is unlikely to be a limiting factor in the regeneration of this species unless it is related to a more complex set of competitive interactions imposed by shade, heterotrophs and the non-mycorrhizal condition of young seedlings.

Antagonistic populations in the rhizosphere

The stunting of seedlings of *E. pilularis* in mature forest soil was suggested by Florence & Crocker (1962) to be due to a microbiological antagonism. Subsequently Evans, Cartwright & White (1967) isolated *Cylindrocarpon destructans* (syn. *radicicola*) from the root surfaces of such plants. From the pure culture of some strains they isolated a phytotoxic compound, nectrolide, which proved to be identical to the antibiotic, brefeldin A 12. This substance was shown to stunt the growth and blacken the roots of *E. pilularis* in sterile culture and is implicated in the regeneration of this eucalypt. *Cylindrocarpon destructans* is generally regarded as a weak parasite whose ecological niche is the rhizosphere (Thornton 1965). It is widespread in moist soils of the temperate zones and is common in acid forest soils, often in the lower A horizon (Domsch & Gams 1972).

In spring 1980 this fungus was baited by transplanting twenty 2-leaf stage seedlings from sterilized vermiculite into the mature forest at Wallaby Creek in Half Canopy and Full Canopy conditions. After 2 months the thirteen surviving seedlings were carefully excavated and their roots thoroughly washed in detergent and distilled water according to the method of Harley & Waid (1955). With the assistance of Mr H. Y. Yip, *Cylindrocarpon destructans* was cultured from 2-mm segments from nine of the remaining thirteen seedlings, and was also recovered from the early washings from the roots. Three types of *Cylindrocarpon* isolate were identified, two being forms of *C. destructans*, the other being an unidentified species. A simple exploratory experiment was performed in which 4 ml of the sterile filtrate from each of the three isolates grown on Czapek-Dox medium was added to freshly germinated sterile seedlings of *E. regnans*. After incubating under 100 lux fluorescent illumination at 22 °C for 2 weeks, a statistically significant retardation of radicle growth was found on the two culture solutions of *C. destructans*. After 11 days a marked deterioration of the stunted roots occurred. Between 75 and 100% of the root tips turned black and up to 50% were deformed. The filtrate from *Cylindrocarpon* sp. did not affect the growth or condition of the root (Fig. 6). The results seem strikingly similar to those described by Evans *et al.* (1967) for *E. pilularis*. Although this effect has not been demonstrated to cause diminished growth under the forest canopy, it may contribute to the generally unthrifty condition of seedlings.

It has been observed repeatedly that fresh soil whether sieved or in intact blocks from the mature forest does not support vigorous growth of *E. regnans* seedlings in the glasshouse for 6–8 weeks. Such seedlings frequently develop purple leaf tips and margins.

The effect seems particularly marked when soils are collected in late summer. To investigate whether micro-organisms were involved, top soil (0–12 cm) was collected from the mature forest in early summer (2 December 1980) and coarse-sieved. One-third was immediately placed in a warm glasshouse (20–45 °C) to incubate for 3 months, one-third was stored in a cold room at 2 °C. The remainder was stored at 2 °C for 6 weeks then incubated for 6 weeks. Additional soil from the same area was collected in later summer (17 January 1981); one half was stored cold and the other half was allowed to incubate. In early autumn (5 March 1981) seedlings which had been raised to the 2-leaf stage in vermiculite were planted in 15-cm diameter plastic pots of the variously treated top soils. After 3 weeks, plants in the fully incubated (3 months) soils were green and healthy with a

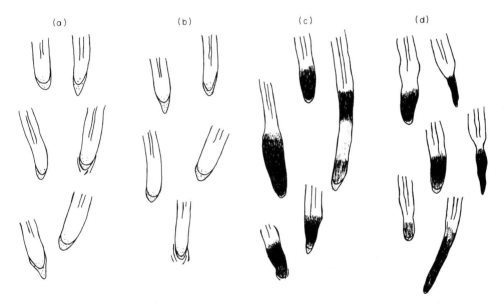

FIG. 6. Sterile seedling radicles of *Eucalyptus regnans* grown in filtrates of *Cylindrocarpon* spp. Much larger differences would have been expected in a less osmotically active medium.

Source of filtrate	(a) Czapek-Dox medium (control)	(b) *Cylindrocarpon* sp.	(c) *C. destructans* (isolate 1)	(d) (isolate 2)
Mean extension/wk (mm)	6·4	8·0	3·9	4·3
Tips blackened (%)	0	0	100	75
Tips deformed (%)	0	0	17	50

mean height of 3·6 cm. Those in the cold-stored soils from the December collection were stunted and purplish and averaged 2·9 cm in height. Most of the plants in the late summer soil collection were also adversely affected, irrespective of any partial incubation. Preliminary analyses after 7 weeks indicated that green plants contained about twice as much phosphorus as purple plants; later analyses are shown in Table 4.

The average mycorrhizal frequency was similar on both green and purple plants at this time, although infection appeared to be more recent on the latter. The roots of the unhealthy plants showed a greater number of dead and dying rootlets and a greater frequency of brown hyphae. After 4 months, most plants in each treatment were associated with fructifications of the mycorrhizal fungus *Laccaria laccata* (Ashton 1976). Partial recovery of seedlings occurred after 8–9 weeks, a period coinciding with full

TABLE 4. Results of a bioassay using *Eucalyptus regnans* seedlings in mature forest top-soils (0–12 cm) incubated moist in the glasshouse for different periods prior to planting. Statistical significance of frequency differences were tested by χ^2, of mean heights, dry weights and phosphorus concentrations by Student's t-test. Within any column the same letter suffixed to each value indicates *non*-significance of the means at $P = 0.05$. Twelve replicates

	Seedling condition at 7 weeks			Seedling condition at 15–18 weeks			
Treatment	Colour % purplish	Height (cm)	% mycorrhizal tips	Colour % purplish 15 wks	Height (cm) 15 wks	Dry wt (g) 18 wks	Phosphorus (μg g^{-1}) 18 wks
Early summer soil collection							
Incubated 12 weeks (20–45 °C)	0 a	3·6 a	76·6	0	18·0 a	2·24 a	745 a
Incubated 6 weeks (20–45 °C)	25 ab	3·0 b	86·0	0	12·7 b	1·17 b	700 a
Not incubated	42 b	2·9 b	81·3	0	10·8 b	0·87 b	574 b
Late summer collection							
Incubated 6 weeks (20–45 °C)	83 c	2·9 b	83·7	0	11·7 b	0·45 c	710 a
Not incubated	75 c	2·3 c	70·2	0	6·6 c	0·55 c	604 b

nitrification of these forest soils in the glasshouse. The height differential between treatments, however, remained until the end of the experiment. At the 6-week-stage of development, five seedlings from each of the five treatments were carefully excavated and their rhizosphere flora investigated by H. Y. Yip in the manner described earlier. From each plant, two first-order roots were selected and divided into six segments each 4-mm long. Each was plated onto malt agar and the resulting colonies identified and counted.

Although considerable variation occurred between each root sample, the pooled data for each seedling showed trends with both soil treatment and seedling health (Table 5). *Cylindrocarpon destructans*, *Mortierella* sp. and bacteria were significantly more frequent on purple seedlings than green, whereas three other fungal species were significantly less frequent. An association analysis of the rhizosphere complement recorded for each seedling was performed using both the monothetic divisive program, DIVINF (Lance & Williams 1968) and the polythetic agglomerative program, MULTBET (Lance & Williams 1967). In each, 5–6 groups were obtained which correlated with seedling health. Amongst the groups identified, the proportion of green healthy seedlings associated with each

TABLE 5. The percentage frequency of the major rhizosphere fungi (and collective bacteria) on the roots of 25 *Eucalyptus regnans* seedlings growing on mature forest soil in the glasshouse. Significance of differences tested by χ^2; *, $P \leqslant 0.05$. Total species $= 66$, number occurring only once $= 30$

		Seedling health		Treatment Incubated (20–45°)	Cold-stored (2°)	Collection 2 Dec.	17 Jan.
Species	Frequency	Green	Purplish				
Penicillium janczewski Zaleski	84	83	85	80	90	80	90
Cylindrocarpon destructans (Zins.) Scholten	76	58 *	92	67 *	90	73	80
P. janthinellum Biourge	52	75 *	31	73 *	20	73	80
P. thomii Maire	40	25	46	67	60	67	40
P. simplicissimum (Oudem.) Thom.	40	58 *	23	53	20	53	20
P. spinulosum Thom.	36	42	31	27	50	33	40
Mortierella ramanniana Möller) Linnem.	36	33	46	27	60	40	40
Trichoderma hamatum (Bon.) Bain	32	50 *	15	27	30	13	20
M. sp. 1	28	8 *	46	13	50	13	50
T. polysporum (Link. ex Pers.) Rifai	24	17	38	13	40	27	30
Paecilomyces carneus (Duché & Heim) Brown & Smith	24	17	31	13	20	13	40
M. alpina Peyron	24	8	38	13	40	13	40
M. nana Linnem.	24	33	15	33	10	33	10
Oidiodendron sp.	20	25	24	20	20	20	30
Dark sterile 1	20	25	15	33	10	33	0
Bacteria (streptomycin-resistant)	36	17 *	54	27 *	70	27	50
Number of seedlings	25	12	13	15	10	15	10

rhizosphere group increased markedly as the ratio of colony numbers of (*Penicillium janthinellum* + *P. simplicissimum*):(*Cylindrocarpon destructans* + bacteria) exceeded unity.

These limited results support the previous suggestion that *C. destructans* may contribute to poor seedling growth under forest conditions, but that its effect may be modified by the presence of other rhizosphere associates.

DISCUSSION AND CONCLUSIONS

The high canopy of mature *E. regnans* forest is unique in that the shade cast is light and hazy. The vigorous understorey in these wet climates is the immediate barrier to successful regeneration of *E. regnans*. Surface fires remove these strata and ameliorate the top-soil by the 'ash-bed' effect (Pryor 1963; Renbuss, Chilvers & Pryor 1972). Under these conditions establishment of *E. regnans* will take place and their crowns will reach the mature canopy if gaps are sufficiently large. Without fire, few seed germinate and very few survive the first year; none remain after 10–12 yr. It is clear that early death is attributable to fungal attack, insect depredations and to the effects of periods of soil moisture deficiency. Plants weakened by shade are particularly prone to such factors.

It would seem that seedlings weakened by allelopathic agents may be predisposed to lethal damage brought about by other environment hazards. Early germination and growth may well be affected by leachates from insect frass or by contact with fresh lipids in the soil. If mycorrhizal fungi are efficient competitors for ammonium, young non-mycorrhizal plants may be disadvantaged. A case can be made for the inhibition of young plants by the exudations of roots and mycorrhizae of the mature trees and mature understorey. Rovira (1969) has highlighted the complexity of the root interface system and the changes in exudation quality and quantity with changing physiological and environmental conditions. The intact root or mycorrhiza is likely to secrete very different materials from that of an excised root system. Newman & Miller (1977) have observed the effects of root exudates on phosphorus uptake, which may be mediated by rhizosphere micro-organisms. As discovered by Florence & Crocker (1962) soil antagonism varies seasonally; a similar situation appears to occur in *E. regnans* forest (Table 4).

The nature of the rhizosphere microflora itself also appears to be important in the health of the developing seedling. *Cylindrocarpon destructans* is implicated as a factor, because of its ubiquity in the mature forest soils, its capacity to act as a weak parasite and its ability to exude a powerful toxin and antibiotic. Glasshouse trials suggest that, if it is a major factor in seedling decline, its effects may be modified markedly by the presence of other micro-organisms. The environmental factors controlling the balance of the rhizosphere flora therefore appear to be of considerable importance in the nutrition and health of seedlings. Seedling phosphorus deficiency appears to have been alleviated by certain incubation treatments in which the microflora was changed. The tendency for *Cylindrocarpon destructans* to favour cooler soils (Thornton 1965) may aggravate this situation under forest canopy.

Therefore gaps, suitable for regeneration, may have to be large enough to allow direct sun to adequately heat the top-soil. Under these conditions a more favourable mycoflora may develop on root systems of seedlings. Gaps larger than 2–3 tree heights (150–250 m) in diameter are likely to provide these conditions. The extensions of fine roots from mature trees is not much greater than $1\frac{1}{2}$ crown radii from the trunk, hence effects of competition and inhibition from this source would be minimized.

The development of gaps of these dimensions in the mature forest are rare and the nature of the light-break produced is determined by the extent of the smash inflicted on the understorey. A single large tree will smash an area about 10–20-m in diameter and since the crown of giant trees is located in the upper quarter of the tree height, little evidence of the gap is present at the site of the tree butt. Storm damage could produce large breaks and local lightning strikes could provide limited damage or set the conditions for an enormous conflagration. However, the opportunities for such local effects are rare and sporadic. The usual consequence of small gap formation is the regeneration of the more shade-tolerant understorey species and an increase in the fern and scrambling-grass (*Tetrarrhena juncea*) stratum.

Regeneration of the large forests therefore is concerned with catastrophes—principally crown fires in Victoria, when huge areas are razed and even-aged stands initiated (Ashton 1980).

The apparent paradox of a successful species with poor resistance to seed harvesting, fungal attack, insect and marsupial depredations as well as auto-inhibition and soil antagonisms must lie in the natural mode of forest propagation. Since these inflammable forests are burnt every 2–4 centuries and regenerate prolifically from seed, miraculously protected by the small woody capsules, selection for resistance to the galaxy of lethal factors is of little consequence (Newman 1978; Smith 1979).

The 'weak link in the chain' under canopy conditions is the young 2-leaf seedling whose physiology and microbial associates appear to render it particularly susceptible to the antagonisms of the forest microclimate. The unravelling of the complexity of interactions has only just begun. Some comfort from the awe of this complexity can be obtained from the thought contained in the paraphrase of T. S. Elliot in the last paragraph of Watt's (1947) treatise on 'Pattern and Process': 'in order to know any of it, we must know all of it'.

ACKNOWLEDGMENTS

The work was commenced at the instigation of Professor J. S. Turner and was initiated with Mr J. H. Chinner. Most of the field work was carried out in water catchment reserves with the kind permission of the Melbourne and Metropolitan Board of Works. We are indebted to Mr H. Y. Yip for technique and fungal identification and to Mr J. Pederick and Mr P. Kristensen for technical assistance. A Melbourne University Research Grant was held by one of us (E.J.W.).

REFERENCES

Al-Mousawi, A. H. & Al-Naib, F. A. G. (1975). Allelopathic effects of *Eucalyptus microtheca* F. Muell. *Journal of the University of Kuwait (Science)*, **2**, 59–66.

Al-Mousawi, A. H. & Al-Naib, F. A. G. (1976). Volatile growth inhibitors produced by *Eucalyptus microtheca. Bulletin of the Biological Research Center (Baghdad)*, **7**, 17–23.

Al-Naib, F. A. G. & Al-Mousawi, A. H. (1976). Allelopathic effects of *Eucalyptus microtheca*. Identification and characterization of the phenolic compounds in *Eucalyptus microtheca. Journal of the University of Kuwait (Science)*, **3**, 83–88.

Ashton, D. H. (1975). Studies on the litter of *Eucalyptus regnans* F. Muell. forests. *Australian Journal of Botany*, **23**, 867–887.

Ashton, D. H. (1976). Studies on the mycorrhizae of *Eucalyptus regnans* F. Muell. *Australian Journal of Botany*, **24**, 723–741.

Ashton, D. H. (1979). Seed harvesting by ants in forests of *Eucalyptus regnans* F. Muell. in central Victoria. *Australian Journal of Ecology*, **4**, 265–277.

Ashton, D. H. (1980). Fire in tall open-forests (wet sclerophyll forests). *Fire and the Australian Biota* (Ed. by A. M. Gill, R. H. Groves & I. R. Noble), pp. 339–366. Australian Academy of Science, Canberra.

Ashton, D. H. & Macauley, B. J. (1972). Winter leaf disease of seedlings of *Eucalyptus regnans* and its relation to forest litter. *Transactions of the British Mycological Society*, **58**, 377–386.

Ashton, D. H. & Turner, J. S. (1979). Studies on the compensation point of *Eucalyptus regnans* F. Muell. *Australian Journal of Botany*, **27**, 589–607.

Baker, R. T. & Smith, H. G. (1920). *A Research on the Eucalypts and their Essential Oils*, 2nd edn. Technical Museum N.S.W., Sydney.

Cremer, K. W. (1962). Effect of fire on seed shed from *Eucalyptus regnans*. *Australian Forestry*, **29**, 251–262.

del Moral, R. & Muller, C. H. (1969). Fog drip: a mechanism of toxin transport from *Eucalyptus globulus*. *Bulletin of the Torrey Botanical Club*, **96**, 467–475.

del Moral, R. & Muller, C. H. (1970). The allelopathic effects of *Eucalyptus camaldulensis*. *The American Midland Naturalist*, **83**, 254–282.

del Moral, R., Willis, R. J. & Ashton, D. H. (1978). Suppression of coastal heath vegetation by *Eucalyptus baxteri*. *Australian Journal of Botany*, **26**, 203–219.

Domsch, D. H. & Gams, W. (eds) (1972). *Fungi in Agricultural Soils*. Longmans, London.

Ellis, R. C. (1969). The respiration of the soil beneath some eucalypt forest stands as related to the productivity of the stands. *Australian Journal of Soil Research*, **7**, 349–358.

Evans, G., Cartwright, J. B. & White, N. H. (1967). The production of a phytoxin, nectrolide, by some root surface isolates of *Cylindrocarpon radicicola*. *Plant and Soil*, **26**, 253–260.

Florence, R. G. & Crocker, R. L. (1962). Analysis of blackbutt (*Eucalyptus pilularis* Sm.) seedling growth in a blackbutt forest soil. *Ecology*, **43**, 670–679.

Häsler, R. C. & Wanner, H. (1977). Gesättige Fettsauren in Boden von Caffeepflanzungen. *Plant and Soil*, **48**, 397–408.

Harley, J. L. & Waid, J. S. (1955). A method of studying active mycelia on living root and other surfaces in soil. *Transactions of the British Mycological Society*, **38**, 104–118.

Hillis, W. E. (1966). Variations in polyphenol composition within species of *Eucalyptus* L'Hérit. *Phytochemistry*, **5**, 541–556.

Hillis, W. E. (1967). Polyphenols in the leaves of *Eucalyptus*: a chemotaxonomical survey. II. The section *Renantheridae* and *Renantherae*. *Phytochemistry*, **6**, 259–274.

Jackson, W. D. (1968). Fire, air, water and earth—an elemental ecology of Tasmania. *Proceedings of the Ecological Society of Australia*, **3**, 9–16.

Lance, G. N. & Williams, W. T. (1967). A general theory of classificatory sorting strategies. *Computer Journal*, **9**, 373–380.

Lance, G. N. & Williams, W. T. (1968). Note on a new information statistic classificatory program. *Computer Journal*, **11**, 195.

Müller, P. E. (1887). *Studien über die natürlichen Humusformen und deren Einwerkungen mit Vegetation und Boden*. Julius Springer, Berlin.

Muller, C. H. (1970). The role of allelopathy in the evolution of vegetation. *Biochemical Coevolution* (Ed. by K. L. Chamber), pp. 13–31. Oregon State University Colloquium No. 29, Oregon State University Press, Corvallis.

Newman, E. I. (1978). Allelopathy: adaptation or accident? *Biochemical Aspects of Plant and Animal Coevolution* (Ed. by J. B. Harborne), pp. 329–342. Academic Press, London.

Newman, E. I. & Miller, M. H. (1977). Allelopathy among some British grassland species. II. Influence of root exudates on phosphorus uptake. *Journal of Ecology*, **65**, 399–412.

Otroshchenko, O. S., Stepanichenko, N. N., Gusakova, S. D. & Mukhamedzhanov, S. Z. (1979). Lipids and secondary metabolites of Fungi Imperfecti causing plant diseases. *Chemistry of Natural Compounds*, **15**, 236–257.

Pryor, L. D. (1963). Ash-bed growth response as a key to plantation establishment on poor sites. *Australian Forestry*, **27**, 48–51.

Renbuss, M. A., Chilvers, G. A. & Pryor, L. D. (1972). Microbiology of an ashbed. *Proceedings of the Linnean Society of N.S.W.*, **97**, 302–310.

Rice, E. L. & Pancholy, S. D. (1972). Inhibition of nitrification by climax ecosystems. *American Journal of Botany*, **59**, 1033–1040.

Richards, P. W. (1952). *The Tropical Rain Forest: an Ecological Study*. Cambridge University Press, Cambridge.

Robinson, R. K. (1971). Importance of soil toxicity in relation to the stability of plant communities. *The Scientific Management of Animal and Plant Communities for Conservation* (Ed. by E. Duffey & A. S. Watt), pp. 105–113. Blackwell Scientific Publications, Oxford.

Romashkevich, I. F. (1964). Role of bitumens in delaying mobilization of nitrogen compounds in peats and uptake of nitrogen by plants. *Soviet Soil Science*, **1964**, 81–84.

Rovira, A. D. (1969). Plant root exudates. *The Botanical Review*, **55**, 35–56.

Shreiner, O., Reed, H. S. & Skinner, J. J. (1907). Certain organic constituents of soil in relation to soil fertility. *U.S. Department of Agriculture Bureau of Soils Bulletin,* **47,** 1–52.

Smith, A. P. (1979). The paradox of autotoxicity in plants. *Evolutionary Theory,* **4,** 173–180.

Taylor, N. H. & Pohlen, I. J. (1970). Soil survey method. *Soil Survey Bulletin,* **25,** 28–48.

Thornton, R. H. (1965). Studies of fungi in pasture soils. I. Fungi associated with live roots. *New Zealand Journal of Agricultural Research,* **8,** 417–447.

Toomey, J. W. & Korstian, C. F. (1947). *Foundations of Sylviculture upon an Ecological Basis.* John Wiley & Sons, New York.

Trenbath, B. R. & Fox, L. R. (1976). Insect frass and leaves from *Eucalyptus bicostata* as germination inhibitors. *Australian Seed Science Newsletter,* **2,** 34–39.

Veblen, T. T. & Ashton, D. H. (1982). The regeneration status of *Fitzroya cupressoides* in the Cordillera Pelada, Chile. *Biological Conservation* (in press).

Ward, S. C. (1979). *Tree-induced patterns of soil chemistry in some Eastern Australian Forests.* B.Sc. (Hons) thesis, University of New England, N.S.W.

Watt, A. S. (1947). Pattern and process in the plant community. *Journal of Ecology,* **35,** 1–22.

Watt, A. S. (1961). Ecology. *Contemporary Botanical Thought* (Ed. by A. M. MacLeod & L. S. Cobley), pp. 115–131. Oliver and Boyd, London.

Willis, E. J. (1980). *Allelopathy and its role in forests of* Eucalyptus regnans *F. Muell.* Ph.D. thesis, University of Melbourne.

Willis, J. H. (1970). *A Handbook to Plants in Victoria.* Vol. 1. Melbourne University Press, Melbourne.

Willis, J. H. (1972). *A Handbook to Plants in Victoria.* Vol. 2. Melbourne University Press, Melbourne.

Zinke, P. J. (1962). The pattern of influence of individual forest trees on soil properties. *Ecology,* **43,** 130–133.